THE ILLUSTRATED FLORA OF ILLINOIS

The Illustrated Flora of Illinois

ROBERT H. MOHLENBROCK, General Editor

ADVISORY BOARD:

Constantine J. Alexopoulos, *University of Texas*

Gerald W. Prescott, *University of Montana*

Aaron J. Sharp, *University of Tennessee*

Robert F. Thorne, *Rancho Santa Ana Botanical Garden*

Rolla M. Tryon, Jr., *The Gray Herbarium*

THE ILLUSTRATED FLORA OF ILLINOIS

FLOWERING PLANTS
nightshades to mistletoe

Robert H. Mohlenbrock

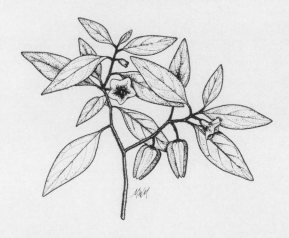

SOUTHERN ILLINOIS UNIVERSITY PRESS
Carbondale and Edwardsville

This book is dedicated to
Trent Alan Mohlenbrock,
who has accompanied me on many
of my field trips.

Copyright © 1990 by the Board of Trustees, Southern Illinois
University
All rights reserved
Printed in the United States of America
Designed by Andor Braun
Production supervised by Natalia Nadraga
93 92 91 90   4 3 2 1

Library of Congress Cataloging-in-Publication Data
Mohlenbrock, Robert H., 1931–
  Flowering plants, nightshades to mistletoe / Robert H.
Mohlenbrock.
    p.   cm.—(The Illustrated flora of Illinois)
  Bibliography: p.
  Includes index.
  ISBN 0-8093-1567-X
    1. Dicotyledons—Illinois—Identification.   2. Dicotyledons—
Illinois—Pictorial works.   3. Dicotyledons—Illinois—Geographi-
cal distribution—Maps.   I. Title.   II. Series.
QK157.M6223   1990
583'.09773—dc20   89-6353
                                                               CIP

CONTENTS

ILLUSTRATIONS

FOREWORD

This is the sixth volume in The Illustrated Flora of Illinois series devoted to dicotyledons, or dicot plants. Five volumes of monocotyledonous plants, or monocots, have been published previously, as well as single volumes on ferns and diatoms.

The concept of The Illustrated Flora of Illinois is to produce a multivolumed flora of the plants of the state of Illinois that will include algae, fungi, mosses, liverworts, lichens, ferns, and seed plants. For each kind of plant known to occur without cultivation in Illinois, a complete description, illustrations showing diagnostic features, distribution maps, and ecological notes will be provided. Keys to aid in identification of the plants will be presented.

An advisory board was created in 1964 to criticize, evaluate, and make suggestions for each volume of The Illustrated Flora of Illinois during its preparation. The board consists of botanists eminent in their area of specialty—Dr. Constantine J. Alexopoulos, University of Texas (fungi); Dr. Gerald W. Prescott, University of Montana (algae); Dr. Aaron J. Sharp, University of Tennessee (mosses, liverworts, lichens); Dr. Robert F. Thorne, Rancho Santa Ana Botanical Garden (flowering plants); and Dr. Rolla M. Tryon, Jr., The Gray Herbarium of Harvard University (ferns).

In volumes of the series not written by me, I shall serve as editor.

There is no definite sequence for publication of The Illustrated Flora of Illinois. Volumes will appear as they are completed.

Robert H. Mohlenbrock

Southern Illinois University
January 30, 1989

THE ILLUSTRATED FLORA OF ILLINOIS

FLOWERING PLANTS
nightshades to mistletoe

WISCONSIN

IOWA

INDIANA

MISSOURI

KENTUCKY

County Map of Illinois

JO DAVIESS
STEPHENSON
WINNEBAGO
BOONE
McHENRY
LAKE
CARROLL
OGLE
De KALB
KANE
COOK
DuPAGE
WHITESIDE
LEE
KENDALL
WILL
HENRY
BUREAU
La SALLE
GRUNDY
ROCK ISLAND
MERCER
STARK
PUTNAM
KANKAKEE
KNOX
MARSHALL
LIVINGSTON
HENDERSON
WARREN
PEORIA
WOODFORD
IROQUOIS
McLEAN
HANCOCK
McDONOUGH
FULTON
TAZEWELL
FORD
MASON
LOGAN
De WITT
CHAMPAIGN
VERMILION
SCHUYLER
PIATT
ADAMS
MENARD
MACON
DOUGLAS
EDGAR
BROWN
CASS
SANGAMON
SCOTT
MORGAN
MOULTRIE
COLES
PIKE
GREENE
MACOUPIN
CHRISTIAN
SHELBY
CUMBERLAND
CLARK
CALHOUN
MONTGOMERY
JERSEY
FAYETTE
EFFINGHAM
JASPER
CRAWFORD
MADISON
BOND
CLAY
RICHLAND
LAWRENCE
CLINTON
MARION
WAYNE
EDWARDS
ST CLAIR
WASHINGTON
JEFFERSON
WABASH
MONROE
RANDOLPH
PERRY
HAMILTON
WHITE
FRANKLIN
JACKSON
SALINE
GALLATIN
WILLIAMSON
UNION
JOHNSON
POPE
HARDIN
ALEXANDER
PULASKI
MASSAC

Introduction

Plants that produce flowers are divided into two groups, the dicotyledons and the monocotyledons. Dicotyledons, or dicots, form two "seed leaves," or cotyledons, when the seed germinates. Monocotyledons, or monocots, form only a single "seed leaf." In the world, as in Illinois, dicots exceed the monocots in number of species. This is the sixth volume of The Illustrated Flora of Illinois that includes dicots. Five volumes on monocots and one each on ferns and diatoms have also been published in this series.

For The Illustrated Flora of Illinois, I have used a modified version of the Thorne system of classification (1968). Thorne's system, utilizing information gathered from cytology, biochemistry, anatomy, and embryology, more clearly depicts natural relationships among flowering plants than does the more familiar Engler and Prantl system used in most floras.

Since the arrangement of orders and families of flowering plants proposed by Thorne is unfamiliar to many, an outline of the orders and families of flowering plants known to occur in Illinois is presented. Those groups in boldface are described in this volume of The Illustrated Flora of Illinois.

Order Annonales
Family Magnoliaceae
Family Annonaceae
Family Calycanthaceae
Family Aristolochiaceae
Family Lauraceae
Family Saururaceae
Order Berberidales
Family Menispermaceae
Family Ranunculaceae
Family Berberidaceae
Family Papaveraceae
Order Nymphaeales
Family Nymphaeaceae
Family Ceratophyllaceae
Order Sarraceniales

Family Sarraceniaceae
Order Theales
Family Aquifoliaceae
Family Hypericaceae[1]
Family Elatinaceae
Family Ericaceae
Order Ebenales
Family Ebenaceae
Family Styracaceae
Family Sapotaceae
Order Primulales
Family Primulaceae
Order Cistales
Family Violaceae
Family Cistaceae
Family Passifloraceae

Family Cucurbitaceae
Family Loasaceae
Order Salicales
Family Salicaceae
Order Tamaricales
Family Tamaricaceae
Order Capparidales
Family Capparidaceae
Family Resedaceae
Family Brassicaceae
Order Malvales
Family Sterculiaceae
Family Tiliaceae
Family Malvaceae
Order Urticales
Family Ulmaceae
Family Moraceae
Family Urticaceae
Order Rhamnales
Family Rhamnaceae
Family Elaeagnaceae
Order Euphorbiales
Family Thymelaeaceae
Family Euphorbiaceae
Order Solanales
Family Solanaceae
Family Convolvulaceae
Family Cuscutaceae[2]
Family Polemoniaceae
Order Campanulales
Family Campanulaceae
Order Santalales
Family Celastraceae
Family Santalaceae
Family Viscaceae
Order Oleales
Family Oleaceae
Order Geraniales
Family Linaceae
Family Zygophyllaceae
Family Oxalidaceae
Family Geraniaceae
Family Balsaminaceae
Family Limnanthaceae

Family Polygalaceae
Order Rutales
Family Rutaceae
Family Simaroubaceae
Family Anacardiaceae
Family Sapindaceae
Family Aceraceae
Family Hippocastanaceae
Family Juglandaceae
Order Myricales
Family Myricaceae
Order Chenopodiales
Family Phytolaccaceae
Family Nyctaginaceae
Family Aizoaceae
Family Cactaceae
Family Portulacaceae
Family Chenopodiaceae
Family Amaranthaceae
Family Caryophyllaceae
Family Polygonaceae
Order Hamamelidales
Family Hamamelidaceae
Family Platanaceae
Order Fagales
Family Fagaceae
Family Betulaceae
Family Corylaceae
Order Rosales
Family Rosaceae
Family Fabaceae
Family Crassulaceae
Family Saxifragaceae
Family Droseraceae
Family Staphyleaceae
Order Myrtales
Family Lythraceae
Family Melastomaceae
Family Onagraceae
Order Gentianales
Family Loganiaceae
Family Rubiaceae
Family Apocynaceae
Family Asclepiadaceae[3]

Family Gentianaceae
Family Menyanthaceae
  Order Bignoniales
Family Bignoniaceae
Family Martyniaceae
Family Scrophulariaceae
Familyo Plantaginaceae
Family Orobanchaceae
Family Lentibulariaceae
Family Acanthaceae
  Order Cornales
Family Vitaceae
Family Nyssaceae
Family Cornaceae
Family Haloragidaceae
Family Hippuridaceae

Family Araliaceae
Family Apiaceae[4]
  Order Dipsacales
Family Caprifoliaceae
Family Adoxaceae
Family Valerianaceae
Family Dipsacaceae
  Order Lamiales
Family Hydrophyllaceae
Family Boraginaceae
Family Verbenaceae
Family Phrymataceae[5]
Family Callitrichaceae
Family Lamiaceae
  Order Asterales
Family Asteraceae

[1]Called Clusiaceae by Thorne (1968).
[2]Included in Convolvulaceae by Thorne (1968).
[3]Included in Apocynaceae by Thorne (1968).
[4]Included in Araliaceae by Thorne (1968).
[5]Included in Verbenaceae by Thorne (1968).

Included in this volume are the orders Solanales, with four families, Campanulales, with one family, and Santalales, with three families.

Since only a small number of dicot families are treated in this book, no general key to the dicot families has been provided. The reader is invited to use my companion book, *Guide to the Vascular Flora of Illinois. Revised and Enlarged Edition* (1986), for keys to all families of flowering plants in Illinois.

Families that are included in the Solanales usually have alternate leaves, actinomorphic flowers, united petals, five stamens, and a superior ovary that is not four-parted. The Lamiales, similar by their united petals, usually have zygomorphic flowers and a four-parted ovary. The Gentianales, also similar by their united petals, usually have opposite leaves. Both the Lamiales and Gentianales will be treated in subsequent volumes of The Illustrated Flora of Illinois. The families of Solanales represented in Illinois and included in this book are the Solanaceae, Convolvulaceae, Cuscutaceae, and Polemoniaceae.

The order Campanulales, consisting of only the family Campanulaceae, have alternate leaves, united petals, and five stamens, but differ from the Solanales by their inferior ovary. Some of them also have zygomorphic flowers.

The order Santalales, at first glance in Illinois, seem to include three very unrelated families—the Celastraceae, which contain self-sufficient autotrophic plants; the Santalaceae, which contain plants that are also dependent on other plants to which they attach themselves to the roots of other species; and the Viscaceae, which obtain their nutrients by parasitizing the branches of other species (in Illinois).

The nomenclature for the species and lesser taxa used in this volume has been arrived at after lengthy study of recent floras and monographs. Synonyms, with complete author citation, that have applied to species in Illinois, are given under each species. A description, while not necessarily intended to be complete, covers the more important features of the species.

The common name, or names, is the one used locally in Illinois. The habitat designation is not always the habitat throughout the range of the species, but only for it in Illinois. The overall range for each species is given from the northeastern to the northwestern extremities, south to the southwestern limit, then eastward to the southeastern limit. The range has been compiled from various sources, including examination of herbarium material and some field studies. A general statement is given concerning the range of each species in Illinois. Dot maps showing county distribution for each taxon are provided. Each dot represents a voucher specimen deposited in some herbarium. There has been no attempt to locate each dot with reference to the actual locality within each county.

The distribution in Illinois has been compiled from extensive field study as well as herbarium study. Herbaria from which specimens have been studied are located at Eastern Illinois University, the Field Museum of Natural History, the Gray Herbarium of Harvard University, the Illinois Natural History Survey, the Illinois State Museum, Knox College, the Missouri Botanical Garden, the Morton Arboretum, the New York Botanical Garden, Southern Illinois University at Carbondale, the United States National Herbarium, and Western Illinois University. In addition, some private collections have been examined. The author is indebted to the curators and staffs of these herbaria for the courtesies extended.

I am deeply grateful to the Gaylord and Dorothy Donnelley Foundation for their generous support that made preparation of this volume possible. I wish to thank Henry and Alice Barkhausen of Jonesboro, Illinois, and the Garden Guild of Winnetka, Illinois, for their support in the publication of this book. I wish to acknowledge the meticulous efforts of Mr. Steve Smith, my editor at the Southern Illinois University Press.

The illustrations for each species, depicting the habit and the distinguishing features, were prepared by my son Mark. My daughter Wendy prepared the maps, while my wife Beverly assisted me in several of the herbaria and typed all the drafts of the manuscript. Without the help of all those individuals and organizations mentioned above, this book would not have been possible.

Descriptions and Illustrations

Order Solanales

The Solanales are represented in Illinois by four families—Solanaceae, Convolvulaceae, Cuscutaceae, and Polemoniaceae. The Cuscutaceae are unique in not possessing chlorophyll. All families of this order in Illinois have alternate leaves (except the Cuscutaceae, which have no leaves), actinomorphic flowers, united petals, five stamens, and a superior ovary. Except for the slightly woody solanaceous genus *Lycium* and one species of *Solanum,* all are herbaceous.

SOLANACEAE–NIGHTSHADE FAMILY

Herbs or less frequently somewhat woody, rarely climbing; leaves usually alternate, simple or compound; inflorescence various; flowers perfect, actinomorphic; sepals 5, united, often persistent on the fruit; petals 5, united into a rotate, campanulate, or funnelform corolla; stamens usually 5, attached to the corolla tube; ovary superior, 2- to 6-locular, with numerous ovules on axile placentae; fruit a capsule or berry.

The Solanaceae are a family found in both temperate and tropical regions of the Old and New World. Botanists recognize about eighty genera and more than three thousand species, including such important plants as the potato (*Solanum tuberosum*), eggplant (*S. melongena*), tomato (*Lycopersicum esculentum*), strawberry tomato (*Physalis* spp.), and red pepper (*Capsicum* spp.), and several ornamentals—butterfly plant (*Schizanthus* spp.), salpiglossis (*Salpiglossis* spp.), petunia (*Petunia* spp.), nierembergia (*Nierembergia* spp.), browallia (*Browallia* spp.), brunfelsia (*Brunfelsia* spp.), and the chinese lantern (*Physalis alkekengi*). In addition, several species in the family have poisonous properties, while others are drug plants—belladonna (*Atropa belladonna*), henbane (*Hyoscyamus* spp.), and stramonium (*Datura* spp.). Tobacco (*Nicotiana* spp.) also belongs to this family.

The Illinois flora contains nine genera of Solanaceae, with only *Solanum* and *Physalis* having native species in Illinois.

7

KEY TO THE GENERA OF Solanaceae IN ILLINOIS

1. Plants woody, at least at base, often climbing or trailing _____ 2
1. Plants herbaceous throughout, not climbing or trailing_____ 3
   2. At least some of the leaves with lobes near the base; flowers in terminal or axillary cymes _____ 1. *Solanum*
   2. None of the leaves lobed near the base; flower solitary in the axil of the leaves _____ 2. *Lycium*
3. Leaves compound (some deeply lobed species of *Solanum* may be sought here)_____ 4
3. Leaves simple, although sometimes deeply lobed _____ 6
   4. Stems prickly _____ 1. *Solanum*
   4. Stems glabrous, appressed-pubescent, or pilose _____ 5
5. Flowers yellow; stems pilose _____ 3. *Lycopersicum*
5. Flowers blue, purple, or white; stems glabrous or appressed-pubescent _____ 1. *Solanum*
   6. Stems and sometimes the leaves prickly _____ 1. *Solanum*
   6. Stems and leaves without prickles _____ 7
7. All leaves sessile, with the uppermost clasping_____ 4. *Hyoscyamus*
7. Some or all the leaves petiolate, the uppermost not clasping (although the uppermost broadly rounded and sessile in some species of *Petunia* and *Nicotiana*) _____ 8
   8. Leaves lobed at least halfway to midvein _____ 1. *Solanum*
   8. Leaves entire or toothed, never lobed halfway to midvein_____ 9
9. Corolla pale blue, campanulate _____ 5. *Nicandra*
9. Corolla white, yellow, purple, or pink (rarely pale blue in some forms of *Petunia*), rotate or funnelform _____ 10
   10. Stamens exserted_____ 1. *Solanum*
   10. Stamens included _____ 11
11. Lobes of corolla with a slender, taillike projection at the tip; fruits prickly _____ 6. *Datura*
11. Lobes of corolla rounded or acute, never with a taillike projection at the tip; fruits not prickly_____ 12
   12. Corolla rotate; fruit a berry _____ 7. *Physalis*
   12. Corolla funnelform; fruit a capsule_____ 13
13. Flowers some shade of yellow; stamens all the same size _____ _____ 8. *Nicotiana*
13. Flowers usually some color other than a shade of yellow; one of the stamens considerably smaller than the others _____ 9. *Petunia*

1. *Solanum* L. –Nightshade

Herbs or sometimes a vine woody at the base (in Illinois); leaves alternate, simple but sometimes very deeply lobed; inflorescence

variable; flowers perfect, actinomorphic; sepals 5, green, united below; corolla rotate, often deeply 5-parted; stamens 5, included;
ovary superior; fruit a berry with many seeds.

There are about two thousand species of *Solanum* found in many
parts of the world but particularly concentrated in tropical America.

KEY TO THE SPECIES OF Solanum IN ILLINOIS

1. Plants prickly _____ 2
1. Plants glabrous, pilose, puberulent, viscid, or stellate-pubescent ____
   _____ 6
   2. Leaves deeply lobed, usually nearly to the midvein; calyx prickly __
      _____ 3
   2. Leaves entire to wavy-toothed to coarsely toothed, never lobed
      nearly to the midvein; calyx without prickles _____ 4
3. Flowers yellow; stem pubescence stellate_____ 1. S. *cornutum*
3. Flowers violet to purple; stem pubescence of simple hairs _____
   _____ 2. S. *heterodoxum*
   4. Stems hirsute, the stellate hairs sessile _____ 3. S. *carolinense*
   4. Stems with short tomentum, the stellate hairs stipitate _____ 5
5. Leaves linear-lanceolate to narrowly oblong, silvery-gray; calyx lobes
   linear _____ 4. S. *elaeagnifolium*
5. Leaves ovate, green; calyx lobes ovate _____ 5. S. *dimidiatum*
   6. Plants woody at base, climbing; berry red _____ 6. S. *dulcamara*
   6. Plants herbaceous, not climbing; berry green or black_____ 7
7. Leaves compound _____ 7. S. *tuberosum*
7. Leaves simple, although sometimes deeply lobed _____ 8
   8. Leaves deeply lobed _____ 8. S. *triflorum*
   8. Leaves entire or repand _____ 9
9. Corolla 6–8 mm broad; plants glabrous or nearly so; berry black _____
   _____ 9. S. *ptycanthum*
9. Corolla 7–11 mm broad; plants viscid; berry green or yellow _____
   _____ 10. S. *sarrachoides*

1. **Solanum cornutum** Lam. Illustr. 2:25. 1794. *Fig. 1.*
   Solanum rostratum Dunal, Hist. Nat. Sol. 234, pl. 24. 1813.

Annual from an elongated root; stems erect, branched, densely covered with stiff yellow prickles and hoary stellate pubescence, to 65
cm tall; leaves oval in outline, 1- to 2-pinnatifid, to 10 cm long, the
lobes usually obtuse, hoary stellate-pubescent throughout, occasionally yellow prickly on the main vein below, the petioles often
yellow prickly; inflorescence racemose, the flowers perfect, 2.0–2.5
cm across, on usually prickly pedicels to 1.5 cm long; calyx urceo-

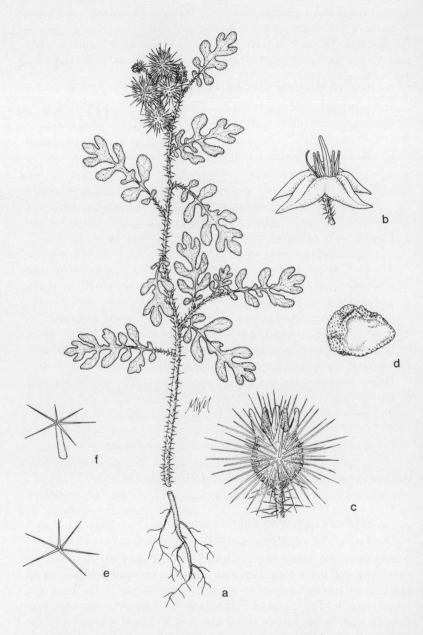

1. Solanum cornutum (Buffalo Bur). *a.* Habit, ×½. *b.* Flower, ×1½. *c.* Fruit, ×1½. *d.* Seed, ×10. *e.* Sessile stellate hair, ×20. *f.* Stalked stellate hair, ×40.

late, green, 5-lobed, the lobes lanceolate, acuminate, densely stellate-pubescent and prickly; corolla rotate, yellow, the 5 lobes broadly ovate, acute; stamens 5, curved, the anther of one of them longer than the others and with an incurved beak; style curved; berry enclosed by the persistent prickly calyx, to 2.5 cm in diameter, with several rugose seeds.

COMMON NAME: Buffalo Bur.

HABITAT: Waste ground, particularly along railroads.

RANGE: Native to the western United States; naturalized in many of the eastern United States.

ILLINOIS DISTRIBUTION: Occasional throughout the state.

The densely yellow-prickly stems, flowers, and fruits of the buffalo bur readily distinguish it from other species of *Solanum* in Illinois. It is also the only yellow-flowered *Solanum* in the state.

In the western states, this plant is a nuisance because the prickly fruits become embedded in sheep's wool. The berry is also reported to be poisonous.

Although this species has been known for many years as *Solanum rostratum,* the earliest legitimate binomial appears to be *S. cornutum.* The similar-appearing but much rarer adventive, *S. heterodoxum,* has purple flowers.

Buffalo bur flowers from July to October.

2. **Solanum heterodoxum** Dunal var. **novomexicanum** Bartl. Proc. Am. Acad. Arts 44:628. 1909. *Fig. 2.*

Androcera neomexicana (Bartl.) Wood & Standl. Contr. U.S. Nat. Herb. 16:170. 1913.

Annual from an elongated taproot; stems erect, branched, densely covered with stiff yellow prickles and pubescence with simple, glandular hairs, to 70 cm tall; leaves ovate to deltoid in outline, 2-pinnatifid, to 11 cm long, the lobes broadly rounded, glandular-pubescent with simple hairs, occasionally yellow prickly on the main vein below, the petioles usually yellow prickly; inflorescence racemose, the flowers perfect, 1.0–1.7 cm across, on usually prickly pedicels to 1.5 cm long; calyx campanulate, green, 5-lobed; corolla rotate, purple, the 5 lobes broadly ovate and usually apiculate; stamens 5, the anther of one of them longer than the others and only slightly incurved; style curved; berry enclosed by the persistent prickly calyx, to 1.2 cm in diameter, with several reticulate seeds.

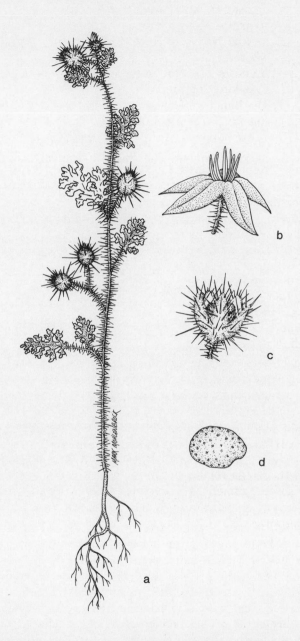

2. *Solanum heterodoxum* var. *novomexicanum* (Purple Horse-nettle). *a.* Habit, × ½. *b.* Flower, × 1½. *c.* Fruit, × 1½. *d.* Seed, × 5.

 COMMON NAME: Purple Horse-nettle.

HABITAT: Along a railroad (in Illinois).

RANGE: Native to New Mexico; rarely adventive else-where.

ILLINOIS DISTRIBUTION: Found only once in 1962 in Crawford County.

The Illinois collection, originally identified as S. *rostra-tum* (= S. *cornutum*) but later verified by Dr. William D'Arcy to be S. *heterodoxum*, differs from S. *cornutum* by its purple flowers. The pubescence on the prickly stem is composed of simple hairs, further matching the pubescence type of S. *heterodoxum*. Additional study of the Illinois specimen proves it to be var. *novomexicanum*.

This rarely adventive plant strongly resembles S. *cornutum*, but is readily distinguished by its purple flowers.

The Illinois specimen was collected in flower on August 11.

3. Solanum carolinense L. Sp. Pl. 184. 1753. *Fig. 3.*

Solanum carolinense L. var. *albiflorum* Kuntze, Rev. Gen. Pl. 2: 454. 1891.

Solanum carolinense L. f. *albiflorum* (Kuntze) Benke, Am. Midl. Nat. 22:213. 1939.

Perennial from a creeping rhizome; stems erect, branched, stellate-hirsute and with short, pale yellowish prickles, to 90 cm tall; leaves oval to oblong, to 15 cm long, shallowly 5- to 9-lobed, the lobes acute, stellate-pubescent, scabrous, occasionally prickly on the veins, the petioles prickly; inflorescence cymose or racemose, the flowers perfect, to 2.5 cm across, on pedicels to 1.5 cm long; calyx rotate, green, 5-lobed, the lobes lanceolate, acute, stellate-pubescent; corolla rotate, purple or white, the 5 lobes narrowly ovate, acute; stamens 5, connivent; berry globose, glabrous, orange-yellow, to 2 cm in diameter, subtended for a while by the subpersistent calyx.

3. Solanum carolinense (Horse-nettle). *a.* Habit, × ½. *b.* Section of stem, × ¼. *c.* Flower, × 1½. *d.* Fruit with calyx, × 1. *e.* Seed, × 5. *f.* Stellate hair, × 10.

COMMON NAME: Horse-nettle.

HABITAT: Waste ground.

RANGE: Massachusetts to Minnesota, south to Texas and Florida.

ILLINOIS DISTRIBUTION: Common; in every county.

This species is particularly common in pastures, fields, and lawns. There is likelihood that it may be adventive in our northernmost counties.

The berries contain a poisonous substance. The prickly stems can cause discomfort when handled.

Flowers may range in color from dark purple to white. White-flowered forms have been designated f. *albiflorum*.

The horse-nettle flowers from June to October.

4. **Solanum elaeagnifolium** Cav. Icon. 3:22, pl. 243. 1794. *Fig. 4.*

Perennial from a creeping rhizome; stems erect, branched, silvery-canescent with stellate hairs, sometimes short prickly, to 75 cm tall; leaves linear-lanceolate to oblong, obtuse to subacute at the apex, tapering to the base, entire to repand-sinuate, to 10 cm long, silvery-canescent and sometimes prickly; inflorescence cymose, the flowers perfect, to 2.5 cm across, on pedicels to 2 cm long; calyx rotate, green, 5-lobed, the lobes linear-lanceolate, acuminate, canescent-tomentose; corolla rotate, bluish, the lobes 5, narrowly ovate, acute; stamens 5; berry globose, glabrous, yellow, to 1.5 cm in diameter, subtended for a while by the subpersistent calyx.

COMMON NAME: Silvery Horse-nettle.

HABITAT: Along railroads.

RANGE: Missouri to Kansas, south to Arizona and Texas; Mexico; adventive in eastern North America.

ILLINOIS DISTRIBUTION: Known from Adams, Cook, Crawford, St. Clair counties.

The silvery horse-nettle is a handsome species with its silvery, stellate-pubescent leaves and stems. There are much fewer prickles on the stem of S. *elaeagnifolium* than on S. *carolinense*.

The flowers occur from July to October.

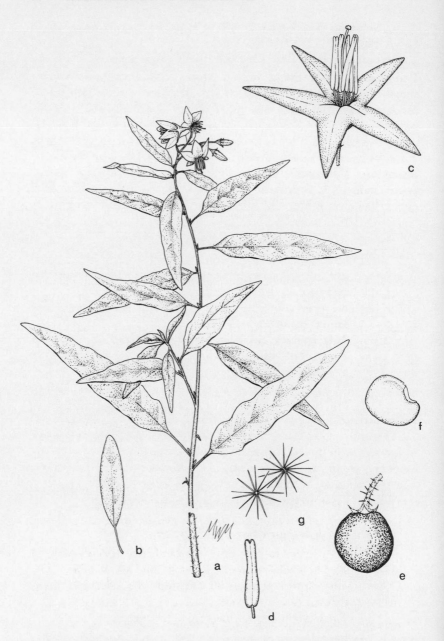

4. Solanum elaeagnifolium (Silvery Horse-nettle). *a.* Habit, ×½. *b.* Leaf, ×¼. *c.* Flower, ×2½. *d.* Anther, ×3. *e.* Fruit, ×1½. *f.* Seed, ×5. *g.* Stellate hairs, ×30.

5. **Solanum dimidiatum** Sendt. in Mart. Fl. Bras. 10:102. 1846.
Fig. 5.
Solanum torreyi Gray, Proc. Am. Acad. 6:44. 1862.

Perennial from a creeping rhizome; stems erect, branched, stellate-tomentose, sometimes prickly, to 1 m tall; leaves ovate in outline, to 15 cm long, shallowly 5- to 9-lobed, the lobes subacute, stellate-tomentose, the petioles canescent and sometimes prickly; inflorescence cymose, the flowers perfect, to 2.5 cm across, on pedicels to 2 cm long; calyx rotate, green, 5-lobed, the lobes broadly ovate, acuminate, tomentose; corolla rotate, violet, the lobes 5, ovate, acute; stamens 5, connivent; berry globose, glabrous, yellow, to 2 cm in diameter, subtended for a while by the subpersistent calyx.

COMMON NAME: Torrey's Horse-nettle.
HABITATS: Waste ground.
RANGE: Arkansas to Kansas, south to Texas; adventive eastward in the United States.
ILLINOIS DISTRIBUTION: Known only from Greene and Henry counties.

This species has leaves similar to those of *S. carolinense*, but differs from this species by its tomentose herbage.

This is another species of the southwestern United States which should be expected along railroads in western Illinois.

This species has been divided sometimes into *S. dimidiatum* and *S. torreyi*, with the Illinois material assignable to *S. torreyi*. Recent evidence seems to indicate that *S. dimidiatum* and *S. torreyi* are one and the same, a treatment followed in this work.

Torrey's horse-nettle flowers from July to September.

6. **Solanum dulcamara** L. Sp. Pl. 185. 1753. *Fig. 6.*
Solanum dulcamara L. var. *villosissimum* Desv. Obs. Pl. Angers 112. 1818.
Solanum dulcamara L. f. *albiflorum* House, Bull. N.Y. State Mus. 254:613. 1824.

Woody perennial; stems climbing, branched, glabrous or villous, to nearly 3 m long; leaves ovate, acuminate at the apex, rounded at the base, often 3-lobed or 3-parted with 2 small basal lobes, glabrous or villous, to 10 cm long, on slender petioles; inflorescence

5. *Solanum dimidiatum* (Torrey's Horse-nettle). *a.* Habit, ×½. *b.* Fruit, ×1. *c.* Stellate hair, ×35.

cymose, the flowers perfect, to 1.5 cm across, on spreading or pendulous, articulated pedicels; calyx rotate, green, 5-lobed, the lobes oblong, obtuse, glabrous or pubescent; corolla 5-lobed nearly to the

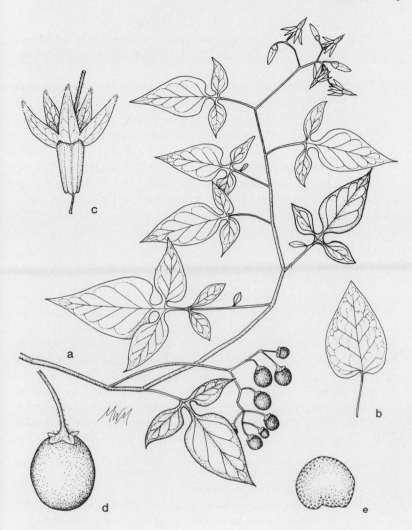

6. *Solanum dulcamara* (Bittersweet Nightshade). *a.* Habit, ×½. *b.* Unlobed leaf, ×½. *c.* Flower, ×3. *d.* Fruit, ×2. *e.* Seed, ×7½.

base, the lobes lanceolate, acute to acuminate, violet to white; stamens 5, connivent; berry globose to ovoid, glabrous, red, to 1.5 cm in diameter, subtended for a while by the subpersistent calyx.

COMMON NAME: Bittersweet Nightshade.

HABITAT: Thickets, marshes, low woods, along fences.

RANGE: Native of Europe; naturalized throughout much of North America.

ILLINOIS DISTRIBUTION: Common in the northern half of the state, occasional in the southern half.

This species, sometimes planted as a garden ornamental, can be found in a wide variety of habitats, from disturbed areas to more "natural" localities.

There is variation in the degree of lobing of the leaves, in flower color ranging from violet to white, and in degree of pubescence of the stems and leaves. White-flowered plants are sometimes designated as f. *albiflorum*. Plants with villous stems and leaves are sometimes called var. *villosissimum*.

The bittersweet nightshade flowers from June to October.

7. **Solanum tuberosum** L. Sp. Pl. 185. 1753. *Fig. 7.*

Herbaceous perennial from slender rhizomes and enlarged tubers; stems decumbent to more or less erect, branched, glabrous or pubescent, to nearly 1 m tall; leaves pinnately compound with 5–9 main leaflets, often with tiny leaflets in between, the leaflets mostly ovate, acute at the apex, tapering to the base, glabrous or pubescent; inflorescence cymose, the flowers bisexual, to 3 cm across, on slender pedicels; calyx rotate, green, 5-lobed, the lobes linear-lanceolate, acute to acuminate; corolla rotate, white to purplish, 5-lobed, the lobes ovate, acute; stamens 5; berry globose, glabrous, yellow or greenish, to 1.8 cm in diameter.

COMMON NAME: Potato.

HABITAT: Waste ground.

RANGE: Native to the Andes of South America; commonly planted but rarely escaping from cultivation.

ILLINOIS DISTRIBUTION: Specimens have been collected in Hancock and Kane counties where they supposedly were persisting.

The common potato of gardens sometimes may grow outside of cultivation, but it rarely persists for more than a year.

The tubers, which are borne at the end of the slender and soon withering rhizomes, provide much food for many people in the World.

The flowers are borne from June to August.

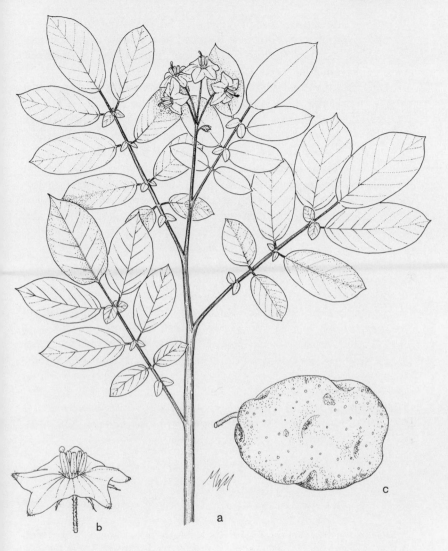

7. *Solanum tuberosum* (Potato). *a*. Habit, × ½. *b*. Flower, × 1½. *c*. Tuber, × ½.

8. Solanum triflorum Nutt. Gen. 1:128. 1818. *Fig. 8.*

Annual from an elongated root; stems erect, branched, pubescent, to 1 m tall; leaves oblong in outline, to 8 cm long, 7- to 11-pinnatifid, the lobes lanceolate, acute, the sinuses rounded, glabrous or pubescent, petiolate; inflorescence 1- to 3-flowered, the

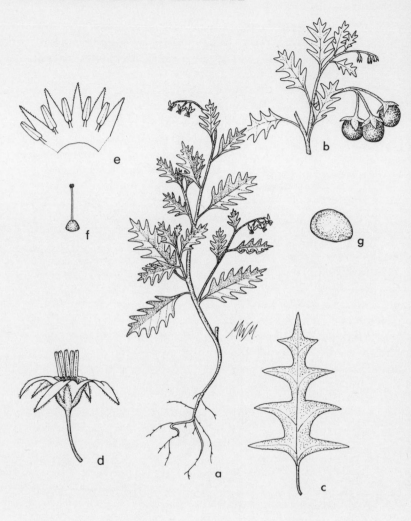

8. *Solanum triflorum* (Cut-leaved Nightshade). *a*. Habit, in flower, ×½. *b*. Habit, in fruit, ×½. *c*. Leaf, ×1½. *d*. Flower, ×3. *e*. Corolla (opened out), with stamens, ×3. *f*. Pistil, ×3. *g*. Seed, ×5.

flowers perfect, to 1 cm across, on reflexed pedicels to 1.5 cm long; calyx 5-lobed, green, the lobes lanceolate, acute; corolla 5-lobed, white, the lobes lanceolate; stamens 5; berry globose, glabrous, green, to 1.5 cm in diameter.

COMMON NAME: Cut-leaved Nightshade.

HABITAT: Waste ground, particularly along railroads.

RANGE: Native of the western United States; adventive eastward.

ILLINOIS DISTRIBUTION: Occasional in the northern counties of the state.

This is another species of *Solanum* that possesses poisonous berries. It is distinguished from all other members of the genus in Illinois by its deeply pinnatifid leaves, its green berries, and its prickleless stems.

This species flowers from June to August.

9. **Solanum ptycanthum** Dunal in DC. Prodr. 13 (1):54. 1852. *Fig.* 9.

Annual from a slender root; stems erect, branched, glabrous or slightly pubescent, to 75 cm long; leaves ovate to lance-ovate, acute to acuminate at the apex, cuneate to rounded at the base, entire to undulate, glabrous or pubescent, to 7.5 cm long, petiolate; inflorescence umbellate, the flowers perfect, to 1 cm across, on slender, reflexed pedicels; calyx 5-lobed, green, the lobes oblong, acute to obtuse; corolla rotate, white or purplish, 5-lobed, the lobes ovate, acute; stamens 5, connivent; berry globose, glabrous, black, lustrous, to 1 cm in diameter.

COMMON NAME: Black Nightshade.

HABITAT: Woodlands, along streams, fields, roadsides, along railroads.

RANGE: Maine to North Dakota, south to Texas and Florida.

ILLINOIS DISTRIBUTION: Common; probably in every county.

Although Illinois plants have been called *S. nigrum* L. or *S. americanum* Mill., neither of these species occurs in Illinois, according to Schilling (1981). He shows that Illinois plants should be known as *S. ptycanthum*. Extremely hairy plants of this species have been found in northern Illinois.

The unripened berries are extremely poisonous.

The flowers, which usually are white but occasionally may be purplish, bloom from May until November.

9. *Solanum ptycanthum* (Black Nightshade). *a.* Habit, with toothed leaves, × ½. *b.* Habit, with entire leaves, × ½. *c.* Flower, × 1½. *d.* Fruit, × 5. *e.* Seed, × 15.

10. **Solanum sarrachoides** Sendtner in Mart. Fl. Bras. 10:18. 1846. *Fig. 10.*

Erect or spreading annual, much-branched, without prickles; leaves ovate, acute at apex, truncate at base, repand, to 12 cm long, to 6 cm broad, viscid; petioles narrowly winged; inflorescence umbellike, few-flowered; calyx 6–8 mm long, the acute lobes about as long as the tube; corolla 7–11 mm broad; berry green or yellow, nearly smooth, 6–8 mm in diameter.

COMMON NAME: Slender Nightshade.
HABITAT: Disturbed soil.
RANGE: Native to the south of Illinois; rarely adventive in the United States.
ILLINOIS DISTRIBUTION: Known from a single collection: St. Clair Co., 4.5 miles north of Highway 50 and County Road 25, on Weil Road, October 25, 1981, *John Berry 44.*
This species, when not in fruit, has more of a resemblance to some species in the genus *Physalis* because of the soft, viscid hairs on the stems and leaves.

Within the genus *Solanum*, this species is nearest to S. *ptycanthum*, but differs by its viscid pubescence, larger flowers, and green or yellow berries.

2. *Lycium* L. –Matrimony Vine

Shrubs or woody vines, often with spines; leaves alternate, simple, entire; flowers perfect, solitary or clustered, axillary or terminal; calyx campanulate, 3- to 5-parted; corolla campanulate, salverform, or funnelform, 5-parted; stamens 5, the anthers dehiscing longitudinally; pistil one, the ovary superior, 2-locular; fruit a berry.

Lycium is a genus of about one hundred species restricted to the warm temperate and subtropical regions of the World. A few species are grown as garden ornamentals.

KEY TO THE SPECIES OF Lycium IN ILLINOIS

1. Leaves gray-green; lobes of calyx obtuse _____ 1. *L. barbarum*
1. Leaves dark green; lobes of calyx acute _____ 2. *L. chinense*

1. Lycium barbarum L. Sp. Pl. 192. 1753. *Fig. 11a–e.*

Lycium halimifolium Mill. Gard. Dict., ed. 8, no. 6. 1768.
Lycium barbarum L. var. *vulgare* Ait. Hort. Kew. 1:257. 1789.
Lycium vulgare (Ait.) Dunal in DC. Prodr. 13 (1):509. 1852.

Shrub, usually without spines; stems recurved, gray, glabrous except for the sometimes slender 1 cm long spines, angled, to 3 m long; leaves lanceolate to oblong, obtuse to acute at the apex, cuneate to the base, entire, gray-green, glabrous, to 6 cm long, the petioles to 1 cm long; flowers axillary in groups of 1–5, to 1.2 cm

10. Solanum sarrachoides (Slender Nightshade). *a.* Habit, × ½. *b.* Fruit, × 1. *c.* Seed, × 10.

broad, on slender pedicels to 2 cm long; calyx campanulate, divided about halfway to the base into usually 3 obtuse lobes, green, glabrous; corolla funnelform, dull purple, with usually 5 ovate-oblong lobes shorter than the tube; stamens 5, the filaments pubescent at the base; berry ovoid to ellipsoid, scarlet to orange-red, 1–2 cm long.

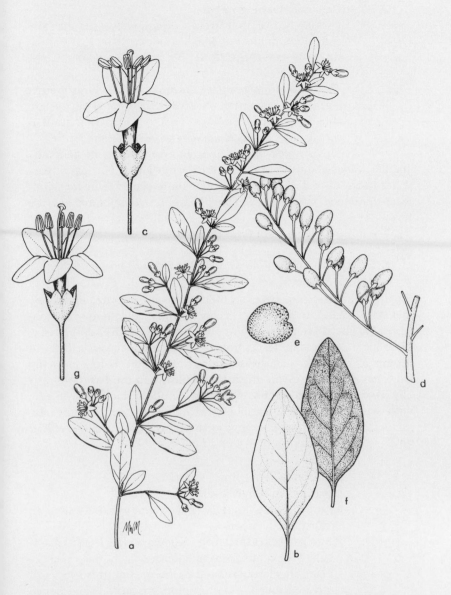

11. *Lycium barbarum* (Common Matrimony Vine). *a.* Habit, ×½. *b.* Leaf, ×1. *c.* Flower, ×2½. *d.* Fruits, ×1. *e.* Seed, ×5. *Lycium chinense* (Chinese Matrimony Vine). *f.* Leaf, ×1. *g.* Flower, ×2½.

COMMON NAME: Common Matrimony Vine.
HABITAT: Waste ground.
RANGE: Native to Europe; occasionally escaped from cultivation in most of North America.
ILLINOIS DISTRIBUTION: Scattered throughout the state.

The common matrimony vine is occasionally planted as a garden ornamental. Its bird-disseminated berries are able to germinate in waste areas so that the plant is occasionally found as an adventive in Illinois.

This species differs from *L. chinense* by its generally smaller leaves and fruits and its obtuse calyx lobes. For many years it has been called *L. halimifolium,* but there seems to be no significant difference between *L. barbarum* and *L. halimifolium,* and the former clearly has priority.

The flowers bloom from June to September.

2. **Lycium chinense** Mill. Gard. Dict., ed. 8, no. 5. 1768. *Fig. 11f, g.*

Shrub without spines; twigs arching or even prostrate, light gray, glabrous, angled, to 4 m long; leaves ovate-lanceolate to ovate, obtuse to acute at the apex, cuneate to the base, entire, dark green, glabrous, to 8 cm long, the petioles to 1 cm long; flowers axillary in groups of 1–5, to 1.2 cm broad, on slender pedicels to 1 cm long; calyx campanulate, divided less than halfway to the base into usually 3–5 acute lobes, green, glabrous; corolla funnelform, purple, with usually 5 ovate-oblong lobes a little longer than the tube; stamens 5, the filaments pubescent at the base; berry ovoid to oblongoid, scarlet to orange-red, 1.5–2.5 cm long.

COMMON NAME: Chinese Matrimony Vine.
HABITAT: Waste ground.
RANGE: Native to Asia; not commonly adventive in eastern North America.
ILLINOIS DISTRIBUTION: Known only from Champaign, Macon, and Mason counties.

Unlike the closely related *L. barbarum,* this species is cultivated much less frequently in gardens and hence escapes less often into waste ground.

The flowers, which have acute calyx lobes, bloom from June to September.

3. *Lycopersicum* Mill.—Tomato

Annual or perennial herbs; leaves alternate, pinnately compound; flowers bisexual, in cymes; calyx 5-lobed; corolla rotate, yellow, 5-lobed; stamens 5, attached to the corolla; pistil 1, the ovary superior, 2- to 3-locular; fruit a berry.

Lycopersicum is a genus of about six species native to South America. It is economically important because of the tomato.

Only the following species occurs in Illinois.

1. **Lycopersicum esculentum** Mill. Gard. Dict., ed. 8, no. 2. 1768. *Fig. 12.*

Solanum lycopersicum L. Sp. Pl. 185. 1753.

Lycopersicon lycopersicon (L.) Karst, Deutsch. Fl. 966. 1882.

Annual or perennial herb from fibrous roots; stems spreading to erect, branched, glandular-pubescent, to 1 m tall; leaves pinnately compound, with 5–9 main leaflets and often several small leaflets formed between some of the main leaflets, to 45 cm long, viscid-pubescent, petiolate, the leaflets oblong to ovate, acute at the apex, coarsely dentate; flowers several in a nodding cyme, to nearly 2 cm broad, on viscid-pubescent peduncles and pedicels; calyx deeply 5-lobed, green, pubescent, the lobes linear-lanceolate, acute to acuminate; corolla rotate, yellow, 5-lobed, the lobes lanceolate to lance-ovate, acute; berry variously shaped, fleshy, red or orange.

COMMON NAME: Tomato.

HABITAT: Waste ground.

RANGE: Native to Tropical America; often planted, occasionally escaped, but rarely persisting from cultivation.

ILLINOIS DISTRIBUTION: Occasional throughout Illinois.

There are many cultivated varieties of the garden tomato. No attempt has been made here to determine which varieties have been found as escapes in Illinois. The flowers appear from May to September.

4. *Hyoscyamus* L. —Henbane

Herbaceous annuals, biennials, or perennials; leaves alternate, toothed or lobed; flowers perfect, large, in axillary clusters; calyx campanulate or urceolate, 5-lobed; corolla funnelform, slightly unequally 5-lobed; stamens 5, free, attached to the corolla; pistil 1,

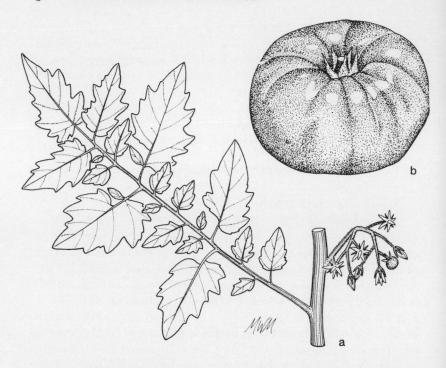

12. *Lycopersicum esculentum* (Tomato). *a.* Leaves and flowers, × ½. *b.* Fruit, × ½.

the ovary superior, 2-locular; capsule 2-locular, circumscissile above the middle, usually enclosed by the calyx.

Hyoscyamus is a genus of about fifteen species mostly native to the Mediterranean region.

Only the following species occurs in Illinois.

1. Hyoscyamus niger L. Sp. Pl. 179. 1753. *Fig. 13.*

Annual or biennial, ill-smelling herbs, from a thickened root; stems erect, branched, viscid-villous, to 75 cm long; leaves oblong in outline, acute to acuminate at the apex, the basal leaves rounded at the petiolate base, the upper leaves cordate-clasping at the sessile base, coarsely sinuate to pinnatifid, viscid-pubescent, to 25 cm long; flowers 1–several in secund spikes, to 4.5 cm across, sessile or nearly so; calyx campanulate, green, villous, 5-lobed, the lobes deltoid,

13. *Hyoscyamus niger* (Black Henbane). *a.* Upper part of plant, × ½. *b.* Flower, × 1¼. *c.* Fruit, × 1½. *d.* Seed, × 5.

subulate-tipped; corolla funnelform, greenish yellow with purple veins, 5-lobed, the lobes ovate, obtuse to subacute; capsule globose to oblongoid, about 1 cm in diameter, enclosed by the enlarged calyx.

COMMON NAME: Black Henbane.
HABITAT: Waste ground.
RANGE: Native of Europe; occasionally naturalized in eastern North America.
ILLINOIS DISTRIBUTION: Known from McHenry and Peoria counties; not collected in Illinois since 1869.
This species is occasionally grown as an ornamental but rarely escapes.
The leaves and inflorescence of the black henbane contain the narcotic hyoscyamin.
The flowers bloom from June to August.

5. *Nicandra* Adans.–Apple-of-Peru

Annual herb; leaves alternate, coarsely toothed or lobed; flowers perfect, solitary in the axils; calyx deeply 5-parted, becoming enlarged and bladdery in fruit; corolla campanulate, the rim nearly entire; stamens 5, free, attached to the corolla; pistil 1, the ovary superior, 3- to 5-locular; berry rather dry, enclosed by the calyx.

Only the following species occurs in Illinois.

1. **Nicandra physalodes** (L.) Gaertn. Fruct. & Sem. 2:237. 1791. *Fig. 14.*
Atropa physalodes L. Sp. Pl. 181. 1753.
Physalodes peruvianum Kuntze, Rev. Gen. Pl. 452. 1891.
Physalodes physalodes (L.) Britt. Mem. Torrey Club 5:287. 1894.

Annual herb; stems erect, branched, angled, glabrous, to 1.5 m tall; leaves oblong to ovate, obtuse to acute at the apex, cuneate to the base, coarsely sinuate, glabrous, to 20 cm long, on glabrous petioles to 3 cm long; flowers solitary in the axils, to 3 cm long, to 3 cm broad, on slender, drooping pedicels; calyx deeply 5-lobed, green, glabrous, the lobes ovate-elliptic, subacute to acute at the apex, sagittate at the base; corolla funnelform, blue, entire or nearly so around the rim; berry globose, to 3 cm across, enclosed in the bladdery calyx.

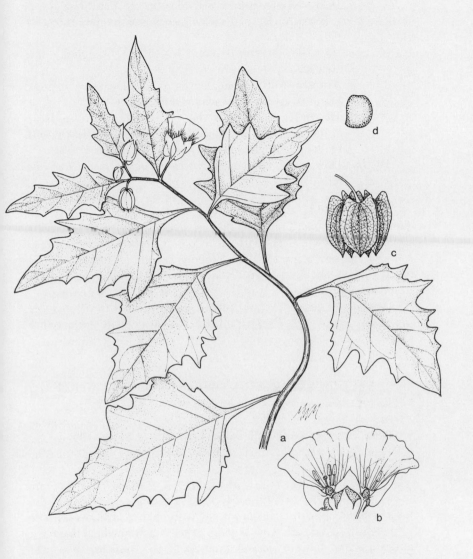

14. Nicandra physalodes (Apple-of-Peru). *a.* Habit, × ½. *b.* Flowers (partly open to show stamens and pistil), × 1½. *c.* Fruit, with calyx, × 1. *d.* Seed, × 5.

COMMON NAME: Apple-of-Peru.

HABITAT: Fields and roadsides.

RANGE: Native of Peru; adventive throughout most of North America.

ILLINOIS DISTRIBUTION: Scattered throughout the state.

A mixture of crushed leaves of this species with milk has been used in the past as a fly poison.

This plant is related to the genus *Physalis* on the basis of the enlarged, bladdery calyx. It differs from *Physalis* by its blue flowers and 3- to 5-celled ovary.

The large flowers bloom from July to September.

6. *Datura* L. –Jimson Weed

Annual or perennial herbs (in Illinois), shrubs, or trees; leaves alternate, usually coarsely toothed or lobed; flowers perfect, large, solitary, axillary; calyx tubular, 5-lobed; corolla funnelform, shallowly 5-lobed; stamens 5, free, attached to the corolla; pistil 1, the ovary superior, 2-locular; fruit a 4-valved, usually prickly capsule.

Datura is a genus of about fifteen species found in most warm regions of the World. Most or all the species contain the narcotic stramonium.

KEY TO THE SPECIES OF Datura IN ILLINOIS

1. Plants glabrous; leaves coarsely toothed; corolla to 10 cm long _____ _____ 1. *D. stramonium*
1. Plants pubescent; leaves entire or undulate; corolla 12–20 cm long ___ _____ 2. *D. innoxia*

1. Datura stramonium L. Sp. Pl. 179. 1753.

Coarse annual herbs; stems erect, stout, branched, glabrous, green or purplish, to 2 m tall; leaves ovate, acute to acuminate at the apex, cuneate at the more or less symmetrical base, sinuate-lobed, glabrous, to 20 cm long, on glabrous petioles to 10 cm long; flower solitary, axillary, to 10 cm long, on pedicels much shorter than the petioles; calyx tubular, angled, less than half as long as the corolla, glabrous, with 5 short, deltoid lobes; capsule ovoid, erect, prickly, to 4.5 cm long.

Two varieties may be distinguished in Illinois.

1. Stem green; corolla white_____ 1a. *D. stramonium* var. *stramonium*
1. Stem purple; corolla purplish _____ 1b. *D. stramonium* var. *tatula*

1a. Datura stramonium L. var. **stramonium** *Fig. 15.*
Stems green; corolla white.

COMMON NAME: Jimson Weed.
HABITAT: Barnyards, fields, waste ground.
RANGE: Native of Asia; naturalized throughout North America.
ILLINOIS DISTRIBUTION: Common; probably in every county.
The name jimson weed is a vulgarization of Jamestown weed, a plant familiar to the early settlers of Virginia. It is a stout, ill-scented plant of farmyards and waste ground.

There is variation in the prickles of the capsule, with longer prickles in the upper part of the capsule and shorter prickles in the lower part.

Most parts of the plant are poisonous, with the seeds especially so.

The flowers bloom from July to October.

1b. Datura stramonium L. var. **tatula** (L.) Torr. Fl. N. & Midl. U.S. 1:232. 1824. *Not illustrated.*
Stems purple; corolla purplish.

COMMON NAME: Purple Jimson Weed.
HABITAT: Barnyards, fields, roadsides.
RANGE: Native to Asia; naturalized throughout much of North America.
ILLINOIS DISTRIBUTION: Occasional in Illinois.
This plant differs from var. *stramonium* by its purplish stems and corollas. The prickles on the capsules of var. *tatula* are generally of a uniform length over the entire capsule.
The flowers bloom from July to October.

2. Datura innoxia Mill. Gard. Dict., ed. 8, Datura no. 5. 1768.
Fig. 16.
Coarse annual herbs; stems erect, stout, branched, glandular-pubescent, green, to 2 m tall; leaves ovate, acute at the apex,

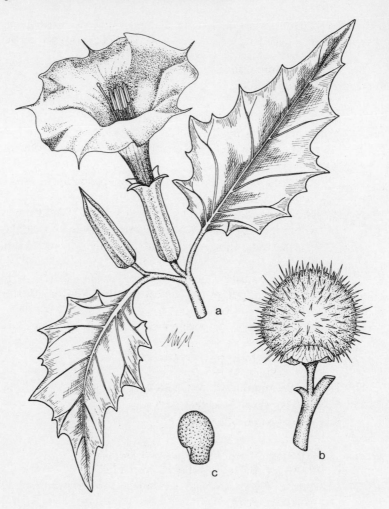

15. *Datura stramonium* (Jimson Weed). *a.* Leaves and flower, × ½. *b.* Fruit, × ½. *c.* Seed, × 5.

rounded at the usually asymmetrical base, entire or undulate, glandular-pubescent, to 25 cm long, on pubescent petioles to 8 cm long; flower solitary, axillary, 12–20 cm long, on pedicels much shorter than the petioles; calyx tubular, not angled, about half as long as the corolla, pubescent, with 5 short, deltoid lobes; capsule globose, nodding, prickly and pubescent, to 3 cm in diameter.

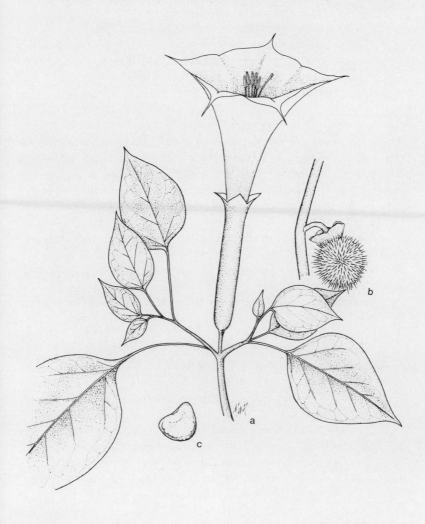

16. *Datura innoxia* (Hairy Jimson Weed). *a*. Leaves and flower, × ½. *b*. Fruit, × ½.
c. Seed, × 1.

COMMON NAME: Hairy Jimson Weed.
HABITAT: Barnyards, waste ground.
RANGE: Native to tropical America; occasionally adventive in the eastern United States.
ILLINOIS DISTRIBUTION: Scattered in Illinois.
This species is very distinct from *D. stramonium* because of its pubescent stems and leaves, nearly entire leaves, and exceptionally large flowers.
Datura innoxia flowers from August to October.

7. *Physalis* L. –Ground Cherry

Annual or perennial herbs; leaves alternate, entire or toothed; flowers perfect, solitary, axillary; calyx 5-lobed, becoming inflated in fruit; corolla yellow or white, often purple-blotched in the center, campanulate to subrotate, 5-lobed; stamens 5, free, attached to the corolla; pistil 1, the ovary superior, 2-locular; fruit a berry, enclosed by the inflated calyx.

Physalis is a genus of about one hundred species native mostly to North and South America. A few species, such as *P. alkekengi*, the Chinese lantern, are grown as garden ornamentals, while *P. pruinosa*, the strawberry tomato, has edible fruits.

The North American species north of Mexico have been studied by Waterfall (1958).

KEY TO THE SPECIES OF Physalis IN ILLINOIS

1. Flowers yellow; calyx during fruiting green or brownish _____ 2
1. Flowers white; calyx during fruiting red or scarlet ___ 15. *P. alkekengi*
   2. Stems glabrous or nearly so, never uniformly villous, hispid, hirsute, or glandular-hairy _____ 3
   2. Stems uniformly villous, hispid, hirsute, or glandular-hairy _____ 9
3. Annuals without rhizomes; leaves mostly conspicuously dentate____ 4
3. Perennials with rhizomes; leaves entire or merely repand-dentate __ 6
   4. Corolla darkened toward the center; peduncles up to 6 mm long; calyx often puberulent, with deltoid teeth; berry purple _____
      _____ 1. *P. ixocarpa*
   4. Corolla without a darkened center; peduncles 10 mm long or longer; calyx usually glabrous, with acute to attenuate teeth; berry yellow
      _____ 5
5. Calyx in fruit strongly 10-angled; peduncles 1–2 cm long _____
      _____ 2. *P. angulata*

5. Calyx in fruit obscurely 10-angled; peduncles nearly all over 2 cm
   long _____ 3. *P. pendula*
   6. Leaves lanceolate to linear; berry yellow _____ 4. *P. longifolia*
   6. Leaves ovate to ovate-oblong; berry red or purple_____ 7
7. Leaves opaque; fruiting calyx ovoid, to 3 cm long_____ 8
7. Leaves translucent; fruiting calyx pyramidal, to 6 cm long _____
   _____ 6. *P. macrophysa*
   8. Leaves tapering to the symmetrical base, up to 6 cm long; peduncle
      to 2 cm long; fruiting calyx sunken at base; plants to 1.5 m tall ____
      _____ 5. *P. subglabrata*
   8. Leaves rounded at the asymmetrical base, up to 4 cm long; peduncle
      to 1 cm long; fruiting calyx not sunken at base; plants less than 1 m
      tall_____ 7. *P. texana*
9. Pubescence of stem glandular-viscid _____ 10
9. Pubescence of stem not glandular-viscid _____ 13
   10. Annuals without rhizomes; fruiting calyx abruptly acuminate__ 11
   10. Perennials with rhizomes; fruiting calyx merely acute _____ 12
11. Fruiting calyx long-tapering, 3–4 cm long; reticulations between the
    lateral nerves obscure_____ 8. *P. barbadensis*
11. Fruiting calyx short-tapering, 2–3 cm long; reticulations between the
    lateral nerves prominent _____ 9. *P. pruinosa*
    12. Leaves more or less tapering to the base, most of them less than 6
        cm long; berry red_____ 10. *P. virginiana*
    12. Leaves more or less rounded or cordate at the base, many of them
        over 6 cm long; berry yellow _____ 11. *P. heterophylla*
13. Peduncles during flowering up to 5 mm long; annuals without rhi-
    zomes_____ 12. *P. pubescens*
13. Peduncles during flowering 1 cm long or longer; perennials with rhi-
    zomes_____ 14
    14. Stems villous; calyx in fruit shallowly or deeply sunken at the base;
        leaves mostly oblong to ovate-lanceolate to ovate_____ 15
    14. Stems hispid or hirsute; calyx in fruit not sunken at the base; leaves
        lanceolate to oblanceolate _____ 14. *P. lanceolata*
15. Leaves more or less tapering to the base _____ 16
15. Leaves more or less rounded or cordate at the base _____
    _____ 11. *P. heterophylla*
    16. Some or all the hairs branched; calyx in fruit mostly 4 cm long or
        longer, shallowly sunken at the base _____ 13. *P. pumila*
    16. All the hairs unbranched; calyx in fruit mostly less than 4 cm long,
        deeply sunken at the base _____ 10. *P. virginiana*

1. **Physalis ixocarpa** Brot. ex Hornem. Hort. Hafn. Suppl. 26. 1816. *Fig. 17.*

Annual from elongated roots; stems spreading to erect, branched, glabrous or with appressed pubescence, to 60 cm tall; leaves ovate to ovate-lanceolate, acute to acuminate at the apex, cuneate to the base, sinuate to dentate to entire, glabrous at maturity, to 7 cm long, on slender, glabrous petioles to 7 cm long; flower solitary, axillary, to 2.5 cm across, on a peduncle to 6 mm long; calyx urceolate, green, sparsely pubescent, obscurely angled, 5-lobed, the lobes deltoid, acute; corolla campanulate, yellow, with a darkened center; berry globose, purple, viscid, enclosed by the inflated calyx, the calyx to 3 cm long.

COMMON NAME: Tomatillo.

HABITAT: Waste ground.

RANGE: Native of the southwestern United States and Mexico; occasionally adventive in eastern North America.

ILLINOIS DISTRIBUTION: Scattered throughout the state, but not common.

This species, along with *P. angulata* and *P. pendula,* are similar in their annual habit, coarsely dentate leaves, and glabrous stems. *Physalis ixocarpa* is distinguished from these two species by its dark-centered corolla, purple berry, and pubescent calyx.

The large yellow flowers and the viscid-purple berry have caused this species to be cultivated on occasion.

The flowers bloom from July to August.

2. **Physalis angulata** L. Sp. Pl. 183. 1753. *Fig. 18.*

Annual from elongated roots; stem erect, branched, glabrous except when very young, to 90 cm tall; leaves ovate to ovate-lanceolate, acute to acuminate at the apex, cuneate to truncate at the base, coarsely dentate to entire, glabrous at maturity, to 10 cm long, on slender, glabrous petioles to 4 cm long; flower solitary, axillary, to 1.3 cm across, on a peduncle 1–2 cm long; calyx urceolate, green, glabrous, strongly 10-angled in fruit, 5-lobed, the lobes lance-attenuate; corolla campanulate, yellow, without a darkened center; berry globose, yellow, enclosed by the inflated calyx to 3.5 cm long.

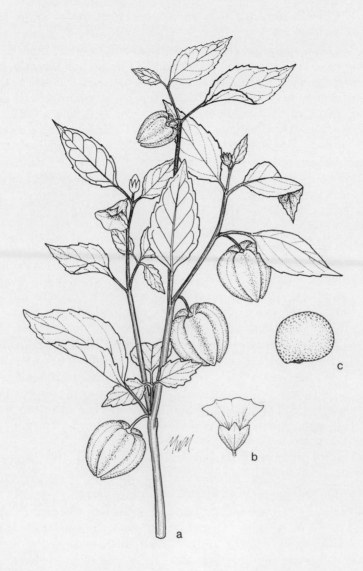

17. *Physalis ixocarpa* (Tomatillo). *a.* Habit, × ½. *b.* Flower, × ¾. *c.* Seed, × 7½.

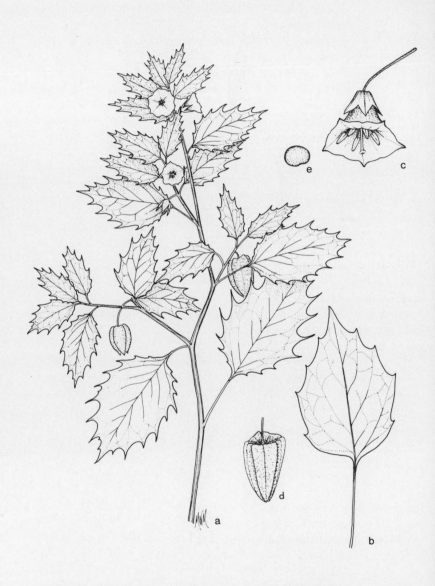

18. Physalis angulata (Ground Cherry). *a.* Habit, × ½. *b.* Leaf, × 1. *c.* Flower, × 2½. *d.* Fruit, × 1. *e.* Seed, × 5.

COMMON NAME: Ground Cherry.

HABITAT: Waste ground.

RANGE: Native to the southern United States; adventive northward.

ILLINOIS DISTRIBUTION: Scattered throughout Illinois, but nowhere common.

This species of the southern United States has been found a few times in Illinois as an adventive in waste ground.

Reports from St. Clair, Cook, and Du Page counties of this species have been referred by Waterfall (1958) to *P.* pendula.

This species flowers from July to September.

3. **Physalis pendula** Rydb. ex Small, Fl. S.E.U.S. 983. 1903. *Fig. 19.*

Physalis angulata L. var. *pendula* (Rydb.) Waterfall, Rhodora 60: 163. 1958.

Annual from elongated roots; stem erect, branched, glabrous except when very young, to 60 cm tall; leaves lanceolate to ovate-lanceolate, acute to acuminate at the apex, cuneate to rounded at the base, coarsely dentate, glabrous at maturity, to 8 cm long, on slender, glabrous petioles to 3 cm long; flower solitary, axillary, to 1 cm across, on peduncles nearly all over 2 cm long; calyx urceolate, green, glabrous, obscurely 10-angled in fruit, 5-lobed, the lobes acute; corolla campanulate, yellow, without a darkened center; berry globose, yellow, enclosed by the inflated calyx, the calyx to 3 cm long.

COMMON NAME: Ground Cherry.

HABITAT: Waste ground.

RANGE: Native to the southwestern United States; adventive in Illinois.

ILLINOIS DISTRIBUTION: Scattered in the southern counties, rare elsewhere.

Waterfall (1958) considers this plant to be a variety of *P. angulata*. The differences exhibited by *P. pendula* are the narrower leaves, longer peduncles, and less attenuated calyx lobes. Until further evidence is available, I am going to maintain specific status for this plant.

The flowers bloom from July to September.

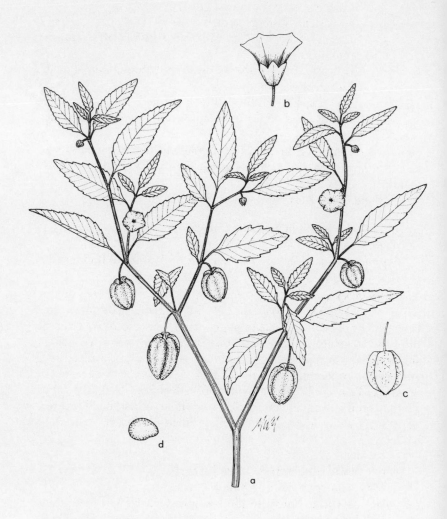

19. Physalis pendula (Ground Cherry). *a.* Habit, ×½. *b.* Flower, ×2½. *c.* Fruit, ×½. *d.* Seed, ×7½.

4. Physalis longifolia Nutt. Trans. Am. Phil. Soc. II, 5:193, 1837. *Fig. 20.*
Physalis pumila Nutt. var. *sonorae* Torr. Bot. Mex. Boundary 153. 1859.
Physalis virginiana Mill. var. *sonorae* (Torr.) Waterfall, Rhodora 60:154. 1958.

20. *Physalis longifolia* (Ground Cherry). *a*. Habit, ×½. *b*. Flower, ×1. *c*. Seed, ×5.

Perennial herb from a stout rhizome; stems erect, branched, angled, glabrous or nearly so, to 1 m tall; leaves lanceolate to linear, subacute to acute at the apex, cuneate to the base, entire, thick, glabrous, to 5 cm long, on glabrous petioles to 3 cm long; flower solitary, axillary, to 2 cm across, on a peduncle to 2 cm long; calyx campanulate, green, glabrous, 5-lobed, the lobes deltoid, about as long as the calyx tube; corolla campanulate, yellow, with a purple center; anthers yellow, tinged with purple; berry globose, yellow, the fruiting calyx ovoid, to 3.5 cm long, scarcely sunken at the base, loosely enclosing the berry.

COMMON NAME: Ground Cherry.

HABITAT: Waste ground.

RANGE: Native to the western United States and Mexico; adventive from Illinois eastward.

ILLINOIS DISTRIBUTION: Known only from Jackson County. A report by Pepoon (1927) from Cook County could not be substantiated.

Although Steyermark (1960) indicates that there is little difference between this and the next taxon (which he calls a variety), differences in Illinois specimens in berry color and leaf shape and texture seem to hold up as indicated in the key. Waterfall (1958) calls this plant *P. virginiana* var. *sonorae.*

The flowers bloom from June to August.

5. **Physalis subglabrata** Mack. & Bush, Trans. Acad. St. Louis 12:86. 1902. *Fig. 21.*

Physalis virginiana Mill. var. *subglabrata* (Mack. & Bush) Waterfall, Rhodora 60:152. 1958.

Physalis longifolia Nutt. var. *subglabrata* (Mack. & Bush) Cronq. Vasc. Pl. Pacif. N.W. 4:286. 1959.

Perennial herb from a stout rhizome; stems erect, branched, angled, glabrous or nearly so, to 1.5 m tall; leaves ovate to ovate-oblong, acute to acuminate at the apex, cuneate to the base, entire or sparsely repand, thin, glabrous or sparsely pubescent, to 6 cm long, on usually glabrous petioles to 6 cm long; flower solitary, axillary, to 2 cm across, on a peduncle to 2 cm long; calyx campanulate, green, glabrous or ciliate, 5-lobed, the lobes deltoid, about as long as the calyx tube; corolla campanulate, yellow, with a purple center; anthers yellow tinged with purple; berry globose, red or purple, the fruiting calyx ovoid, to 3 cm long, sunken at the base, tightly enclosing the berry.

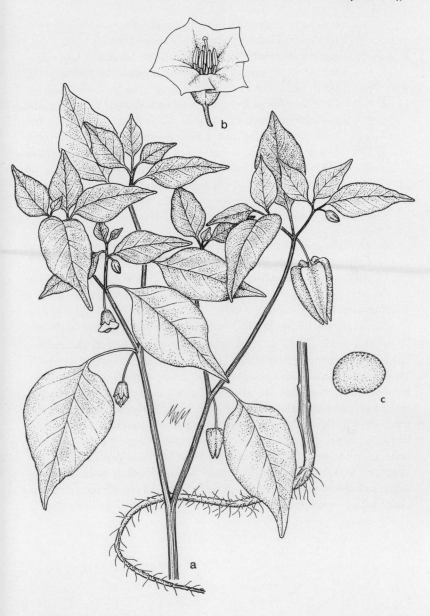

21. *Physalis subglabrata* (Smooth Ground Cherry). *a.* Habit, × ½. *b.* Flower, × 1½. *c.* Seed, × 7½.

COMMON NAME: Smooth Ground Cherry.
HABITAT: Pastures, fields, disturbed woods.
RANGE: Vermont to Iowa, south to Texas and Georgia.
ILLINOIS DISTRIBUTION: Occasional throughout the state.
Considerable differences of opinion exist regarding the correct status of this taxon. I am regarding it as a distinct species for the present time. Cronquist (1959) considers it to be a variety of *P. longifolia*, while Waterfall (1958) calls it a variety of *P. virginiana*. It appears to me to be more closely related to *P. longifolia* than to *P. virginiana*.

The flowers bloom from June to September.

6. Physalis macrophysa Rydb. Bull. Torrey Club 22:308. 1895.
Fig. 22.

Physalis virginiana Mill. f. *macrophysa* (Rydb.) Waterfall, Rhodora 60:153. 1958.

Physalis longifolia Nutt. var. *subglabrata* (Mack. & Bush) Steyerm. f. *macrophysa* (Rydb.) Steyerm. Rhodora 62:131. 1960.

Perennial herb from a stout rhizome; stems erect, branched, angled, glabrous or nearly so, to 1 m tall; leaves ovate, acute to acuminate at the apex, cuneate to the base, coarsely dentate, thin, translucent to transmitted light, glabrous, to 7 cm long, on usually glabrous petioles to 4 cm long; flower solitary, axillary, to 2 cm across, on a peduncle to 2 cm long; calyx campanulate, green, glabrous, 5-lobed, the lobes lanceolate, a little shorter than the calyx tube; corolla campanulate, yellow, with a purple center; anthers yellow, sometimes tinged with purple; berry globose, red or purple, the fruiting calyx pyramidal, 3–6 cm long, deeply sunken at the base, loosely enclosing the berry.

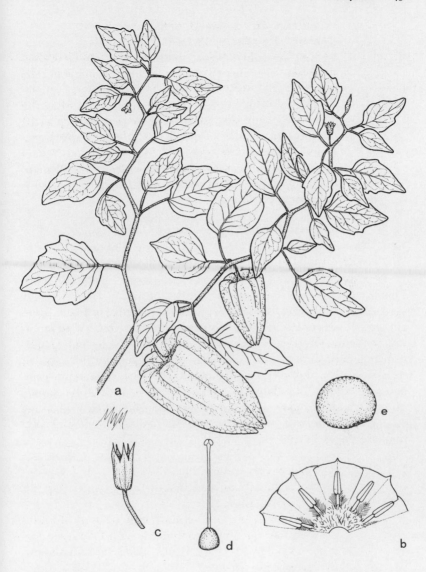

22. *Physalis macrophysa* (Large-fruited Ground Cherry). *a.* Habit, × ½. *b.* Corolla (opened out to show stamens), × 1½. *c.* Calyx, × 1. *d.* Pistil, × 2½. *e.* Seed, × 7½.

COMMON NAME: Large-fruited Ground Cherry.
HABITAT: Usually in low woods.
RANGE: New Jersey to Nebraska, south to Texas and Arkansas.
ILLINOIS DISTRIBUTION: Known from Champaign and Peoria counties.
This rare plant, treated variously as a species, variety, or even form, differs from *P. subglabrata* by its translucent leaves and its large, pyramidal fruiting calyces. I prefer to use the rank of species until more sufficient evidence to the contrary is presented.

The flowers appear in June and July.

7. **Physalis texana** Rydb. Mem. Torrey Club 4:339. 1896. *Fig. 23.*

Physalis virginiana Mill. var. *texana* (Rydb.) Waterfall, Rhodora 60:153. 1958.

Perennial herb from an elongated root; stems low, more or less spreading to suberect, smooth or nearly so, angular, striate, to 30 cm long; leaves ovate, acute to obtuse at the tip, rounded or tapering at the symmetrical or asymmetrical base, usually entire, glabrous on both surfaces, to 4 cm long, on decurrent petioles; flower solitary, axillary, borne on a peduncle up to 1 cm long; calyx campanulate, 5-lobed, the lobes ovate, about as long as the tube; corolla up to 2 cm long, yellow, with a dark center; anthers yellow; fruiting calyx up to 3 cm long, ovoid, more or less 10-angled, not sunken at the base; berry purple.

COMMON NAME: Texas Ground Cherry.
HABITAT: Riverbanks (in Illinois).
RANGE: Texas; Illinois.
ILLINOIS DISTRIBUTION: Known only from a collection made by George Engelmann in St. Clair County in August, 1841, from sandy banks along the Mississippi River, opposite St. Louis.
The Engelmann collection from Illinois is in the herbarium of the Missouri Botanical Garden. It was annotated by P. A. Rydberg during Rydberg's study of the genus in preparation of his published work in 1896. It is interesting to note that although Rydberg annotated the Illinois collection as *P. texana,* he failed to cite the specimen in his revision.

23. *Physalis texana* (Texas Ground Cherry). *a.* Habit, × ½. *b.* Corolla (opened up to show stamens), × 1. *c.* Calyx, × 1. *d.* Pistil, × 1. *e.* Seed, ×5.

The Engelmann specimen matches well the type collection by Heller and a collection by Lindheimer (cited by Rydberg) from Texas.

Waterfall (1958), who also does not attribute this taxon to Illinois, calls this plant *P. virginiana* var. *texana.*

It flowers during the early summer.

8. **Physalis barbadensis** Jacq. Misc. 2:359. 1781. *Fig. 24.*
Physalis obscura Michx. Fl. Bor. Am. 1:149. 1803.
Physalis obscura Michx. var. *glabra* Michx. Fl. Bor. Am. 1:149. 1803.
Physalis barbadensis Jacq. var. *obscura* (Michx.) Rydb. Mem. Torrey Club 4:327. 1896.
Physalis barbadensis Jacq. var. *glabra* (Michx.) Fern. Rhodora 51:82. 1949.

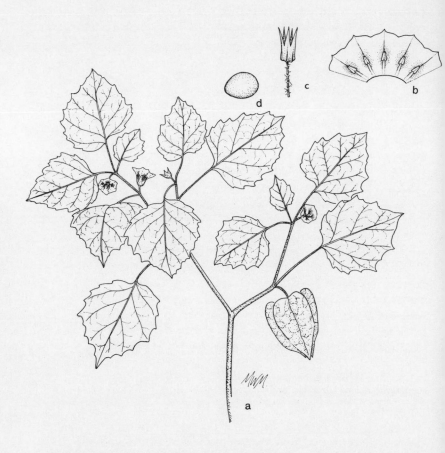

24. *Physalis barbadensis* (Ground Cherry). *a.* Habit, × ½. *b.* Corolla (opened out to show stamens), × 1. *c.* Calyx, × 1½. *d.* Seed, × 5.

Annual from elongated roots; stems spreading to erect, branched, sharply angled, villous and often viscid, rarely glabrous, to 1 m tall; leaves ovate, acute to acuminate at the apex, rounded to subcordate at the base, coarsely dentate, short-pubescent or rarely glabrous, firm, to 5 cm long, on usually pubescent petioles to 4 cm long; flower solitary, axillary, to 1.2 cm across, on a usually pubescent peduncle to 5 mm long, elongating in fruit; calyx campanulate, green, viscid-pubescent, 5-lobed, the lobes lanceolate, as long as or longer than the calyx tube; corolla campanulate, yellow, with a purple center; anthers purple; berry globose, yellow, the fruiting calyx broad below, long-tapering above, to 4 cm long, obscurely reticulate, sunken at the base.

COMMON NAME: Ground Cherry.
HABITAT: Waste ground.
RANGE: Native to tropical America; adventive in the eastern United States.
ILLINOIS DISTRIBUTION: Known only from Jackson County. It has been collected as recently as 1981 from Riverside Park in Murphysboro.
The relationship of this species is with *P. pruinosa*, which has a shorter, less tapering fruiting calyx, and with *P. pubescens*, which lacks the viscidity-pubescence of the stems.

Waterfall (1958) considers the correct name for this plant to be *P. pubescens*. He uses the trinomial *P. pubescens* var. *integrifolia* (Dunal) Waterfall for what most workers in the past have called *P. pubescens*. Kartesz and Kartesz (1980) consider *P. barbadensis* to be a synonym for *P. pubescens* var. *pubescens*.

The flask-shaped fruiting calyx is attractive.

Physalis barbadensis flowers from June to September.

9. **Physalis pruinosa** L. Sp. Pl. 184. 1753. *Fig. 25.*
Physalis pubescens L. var. *grisea* Waterfall, Rhodora 60:167. 1958.

Annual from elongated roots; stems erect, branched, angled, densely villous as well as short-pubescent with viscid hairs, to 60 cm tall; leaves ovate, acute at the apex, cordate at the base, sinuate-toothed, densely short-pubescent with gray, often glandular hairs, firm, to 10 cm long, on pubescent petioles to 4 cm long; flower solitary, axillary, to 1 cm across, on a pubescent peduncle to 5 mm long, elongating in fruit; calyx campanulate, green, villous and often viscid, 5-lobed, the lobes lanceolate, as long as the tube of the calyx; corolla campanulate, yellow, with a purple center; anthers yellow tinged with purple; berry globose, yellow, the fruiting calyx ovoid, short-tapering above, to 3 cm long, prominently reticulate, sunken at the base.

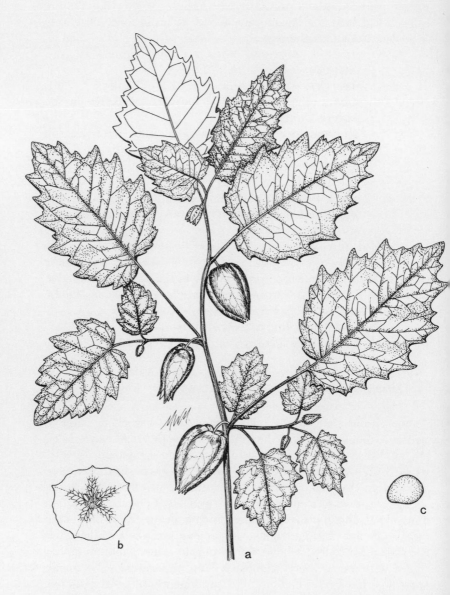

25. *Physalis pruinosa* (Ground Cherry). *a*. Habit, ×½. *b*. Corolla, ×1. *c*. Seed, ×5.

COMMON NAME: Ground Cherry.
HABITAT: Waste ground.
RANGE: Maine to Kansas, south to Texas and Florida.
ILLINOIS DISTRIBUTION: Scattered throughout the state.
This species is considered by Waterfall (1958) to be a variety of *P. pubescens*. It differs from *P. pubescens* by its more cordate leaves.
This ground cherry flowers from June to October.

10. **Physalis virginiana** Mill. Gard. Dict., ed. 8, no. 4. 1768.
Fig. 26.

Perennial herb from deep rhizomes; stems erect, branched, strigose and with some glandular hairs, to 1 m tall; leaves ovate to ovate-lanceolate, subacute to acute at the apex, cuneate to the base, sinuate-dentate to sometimes nearly entire, more or less pubescent, thin, to 5 cm long, on pubescent petioles to 2 cm long; flower solitary, axillary, to 2.5 cm across, on a peduncle to 2 cm long; calyx campanulate, green, pubescent, 5-lobed, the lobes deltoid, about as long as the calyx tube; corolla yellow, with a purple center; anthers yellow; berry globose, red, the fruiting calyx ovoid, 5-angled, pubescent, deeply sunken at the base, up to 3.5 cm long.

COMMON NAME: Ground Cherry.
HABITAT: Woods, disturbed areas.
RANGE: Connecticut to Minnesota, south to Texas and Florida; southern Canada.
ILLINOIS DISTRIBUTION: Occasional throughout the state.
This is one of the more frequently occurring species of *Physalis* in Illinois. It occupies disturbed woodlands, primarily.
This species is quite variable in degree of pubescence.
It is similar to *P. heterophylla* from which it differs by its cordate leaves and its yellow berries.
Flowering time for this species is May to August.

11. **Physalis heterophylla** Nees, Linnaea 6:463. 1831.
Perennial herb from deep rhizomes; stems erect to spreading, usually branched, with glandular pubescence and sometimes some nonglandular villi, to 1 m tall; leaves ovate, acute at the apex,

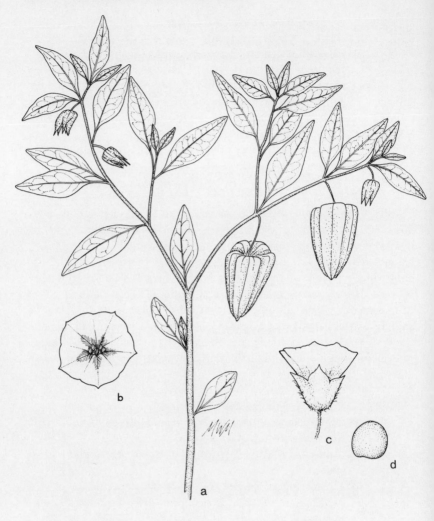

26. Physalis virginiana (Ground Cherry). *a.* Habit, ×½. *b.* Corolla, ×1. *c.* Flower, ×1. *d.* Seed, ×5.

rounded to cordate at the base, dentate to nearly entire, pubescent, thin to firm, to 8 cm long, on pubescent petioles to 3 cm long; flower solitary, axillary, to 2.5 cm across, on a peduncle to 2 cm long; calyx campanulate, green, pubescent, 5-lobed, the lobes deltoid, usually slightly shorter than the calyx tube; corolla yellow, with a purple center; anthers yellow; berry globose, yellow, the fruiting calyx ovoid, to 3 cm long, pubescent, sunken at the base.

Three varieties can usually be distinguished in Illinois.

KEY TO THE VARIETIES OF Physalis heterophylla IN ILLINOIS

1. At least some of the hairs on the stems glandular _____
_____ 11a. *P. heterophylla* var. *heterophylla*
1. Stems without glandular hairs _____ 2
    2. Leaves thick, dentate _____ 11b. *P. heterophylla* var. *ambigua*
    2. Leaves thin, usually entire _____
_____ 11c. *P. heterophylla* var. *nyctaginea*

11a. Physalis heterophylla Nees var. **heterophylla** *Fig. 27-a–d.*
At least some of the hairs on the stem glandular.

COMMON NAME: Ground Cherry.
HABITAT: Woods and disturbed areas.
RANGE: Quebec to Saskatchewan, south to Texas and South Carolina.
ILLINOIS DISTRIBUTION: Scattered throughout the state.
This variety and var. *ambigua* are about equally common in Illinois. The flowers appear from May to September.

11b. Physalis heterophylla Nees. var. **ambigua** (Gray) Rydb. Mem. Torrey Club 4:349. 1896. *Fig. 27e.*
Physalis virginiana Mill. var. *ambigua* Gray, Proc. Am. Acad. 10:65. 1875.
Stems without glandular hairs; leaves thick, dentate.

COMMON NAME: Ground Cherry.
HABITAT: Waste ground.
RANGE: Quebec to Minnesota, south to Texas and Georgia.
ILLINOIS DISTRIBUTION: Occasional throughout the state.
The epithet *ambigua* indicates the doubtful nature of this variety. Waterfall (1958) and others unite this variety with var. *heterophylla*.
The flowers bloom from May to September.

27. *Physalis heterophylla* (Ground Cherry). *a.* Habit, × ½. *b.* Corolla, × 2. *c.* Developing fruit, × ½. *d.* Seed, × 7½. var. *ambigua* (Ground Cherry). *e.* Leaf, × ½. var. *nyctaginea* (Ground Cherry). *f.* Habit, × ½.

11c. Physalis heterophylla Nees var. **nyctaginea** (Dunal) Rydb.
   Mem. Torrey Club 4:349. 1896. *Fig. 27f.*
   Physalis nyctaginea Dunal in DC. Prodr. 13 (1):440. 1852.
   Stems without glandular hairs; leaves thin, usually entire.

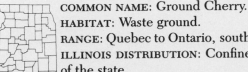

COMMON NAME: Ground Cherry.
HABITAT: Waste ground.
RANGE: Quebec to Ontario, south to Texas and Virginia.
ILLINOIS DISTRIBUTION: Confined to the southern half of the state.
This is the least common variety of *P. heterophylla* in Illinois and possibly the most easily distinguished because of the lack of teeth along the margins of the leaves.
Waterfall (1958) combines this variety with *P. heterophylla* var. *heterophylla*.
The flowers are borne from May to September.

12. **Physalis pubescens** L. Sp. Pl. 183. 1753. *Fig. 28.*
Physalis hirsuta Dunal var. *integrifolia* Dunal in DC. Prodr. 13(1):445. 1852.
Physalis pubescens L. var. *integrifolia* (Dunal) Waterfall, Rhodora 60:166. 1958.

Annuals from elongated roots; stems spreading to ascending, branched, angled, villous, without glandular hairs, to 60 cm long; leaves ovate, acute to acuminate at the apex, rounded to subcordate at the asymmetrical base, entire or sparingly dentate, pubescent at least along the veins, thin, translucent, to 5 cm long, on pubescent petioles to 3 cm long; flower solitary, axillary, to 1 cm across, on a pubescent peduncle to 5 mm long; calyx campanulate, green, pubescent, 5-lobed, the lobes narrowly lanceolate, about as long as the calyx tube; corolla campanulate, yellow, with a purple center; anthers purple; berry globose, yellow, the fruiting calyx ovoid-pyramidal, reticulate, pubescent, sunken at the base, to 3 cm long.

COMMON NAME: Annual Ground Cherry.
HABITAT: Disturbed areas.
RANGE: Virginia to Kansas, south to Texas and Florida; California; Mexico.
ILLINOIS DISTRIBUTION: Occasional throughout the state.
Waterfall (1958) considers this taxon to represent only a variety, calling it *P. pubescens* var. *integrifolia*. He considers the binomial *P. pubescens* L. to be the correct name for what I am calling *P. barbadensis* Jacq. in this work.

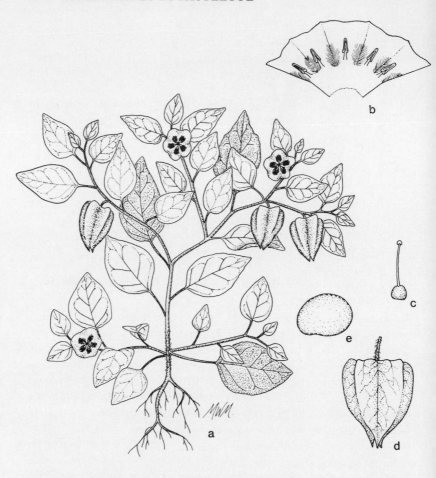

28. *Physalis pubescens* (Annual Ground Cherry). *a.* Habit, × ½. *b.* Corolla (opened out to show stamens), × 2. *c.* Pistil, × 2½. *d.* Fruit, × 1. *e.* Seed, × 15.

This is a very weak-stemmed annual that is often sprawling and very easily crushed.

The flowers are borne from June to October.

13. Physalis pumila Nutt. Trans. Am. Phil. Soc. II, 5:193. 1836.
    Fig. 29.

Perennial herbs from creeping rhizomes; stems erect, branched, angled, to 45 cm tall, villous or hirsute, the hairs often branched;

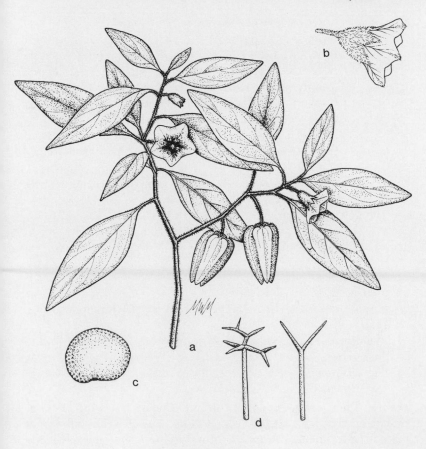

29. *Physalis pumila* (Dwarf Ground Cherry). *a.* Habit, × ½. *b.* Flower, × 1. *c.* Seed, × 7½. *d.* Hair types, × 10.

leaves oblong to ovate, obtuse to acute at the apex, cuneate to the base, entire or rarely sinuate, with forked strigose pubescence, thick, to 9 cm long, on pubescent petioles to 3 cm long; flower solitary, axillary, to 2 cm across, on a pubescent peduncle to 3 cm long; calyx campanulate, green, hirsute, 5-lobed, the lobes deltoid, slightly shorter than the calyx tube; corolla campanulate, yellow, with a brown center; berry globose, yellow, the fruiting calyx ovoid, pubescent, to 4 cm long, slightly sunken at the base.

COMMON NAME: Dwarf Ground Cherry.
HABITAT: Dry hillside.
RANGE: Central Illinois to Colorado, south to Texas.
ILLINOIS DISTRIBUTION: Known only from Peoria County (above Horseshoe Bottom, June 7, 1921, V. H. Chase 3570). The report by Pepoon (1927) from Cook County could not be substantiated.

This species seemingly is native in Peoria County. If this is the case, it is the easternmost natural station for it in the country.

Physalis pumila differs from similar-appearing species by its forked pubescence.

The flowers are borne from June to August.

14. **Physalis lanceolata** Michx. Fl. Bor. Am. 1:149. 1803. *Fig. 30.*

Physalis pennsylvanica Willd. var. *lanceolata* (Michx.) Gray, Man. Bot., ed. 5, 382. 1867.

Physalis virginiana Mill. var. *hispida* Waterfall, Rhodora 60:154. 1958.

Physalis longifolia Nutt. var. *hispida* (Waterfall) Steyerm. Rhodora 62:131. 1960.

Physalis pumila Nutt. ssp. *hispida* (Waterfall) Hinton, Syst. Bot. 1(2):188. 1976.

Perennial herbs from creeping rhizomes; stems spreading to erect, branched, obscurely angled, hispid to hirsute, to 40 cm tall; leaves lanceolate to oblong, obtuse to acute at the apex, cuneate at the base, entire, sparsely short-hairy, thick, to 6 cm long, on pubescent petioles to 2 cm long; flower solitary, axillary, to 1.8 cm across, on a pubescent peduncle to 2 cm long; calyx campanulate, green, pubescent, 5-lobed, the lobes lanceolate, about as long as the calyx tube; corolla campanulate, yellow, with a brown center; berry globose, yellow or green, rarely red, the fruiting calyx ovoid, pubescent, to 3.5 cm long, not sunken at the base.

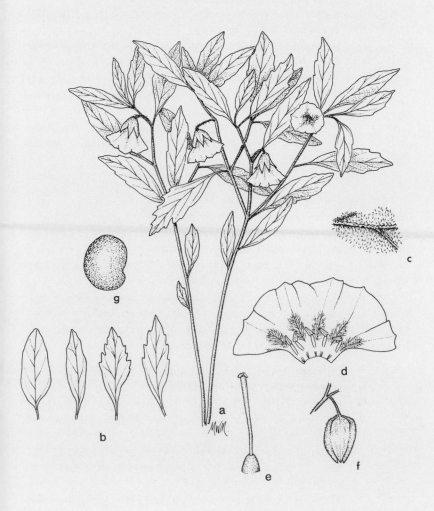

30. Physalis lanceolata (Narrow-leaved Ground Cherry). *a.* Habit, × ½. *b.* Leaf variations, × ½. *c.* Lower surface of leaf, × 10. *d.* Corolla (opened out to show stamens), × 1½. *e.* Pistil, × 2½. *f.* Fruit, × ½. *g.* Seed, × 7½.

COMMON NAME: Narrow-leaved Ground Cherry.

HABITAT: Dry waste ground.

RANGE: Native to the western United States; adventive in eastern North America.

ILLINOIS DISTRIBUTION: Scattered throughout the state, but not common.

Waterfall (1958) prefers to treat this taxon as a variety of *P. virginiana,* while Steyermark (1960) calls it a variety of *P. longifolia.*

The low stature of this plant resembles that of *P. pumila,* but the pubescence of *P. longifolia* is of simple hairs.

The flowers are borne from May to September.

15. Physalis alkekengi L. Sp. Pl. 182. 1753. *Fig. 31.*

Annual or perennial herbs from creeping underground rootstocks; stems flexuous, simple, angled, glabrous or more rarely pubescent, to 75 cm long; leaves ovate, short-acuminate at the apex, rounded at the base, entire or angular, ciliate, to 7 cm long, on usually glabrous, broadened petioles to 2 cm long; flower solitary, axillary, to 1.8 cm across; calyx campanulate, green at anthesis, more or less glabrous, 5-lobed; corolla deeply 5-lobed, white; anthers yellow; berry globose, red, the fruiting calyx ovoid, acuminate, bright red, to 4.5 cm long, deeply sunken at the base.

COMMON NAME: Chinese Lantern.

HABITAT: Persistent near gardens.

RANGE: Native of Asia; rarely escaped from cultivation.

ILLINOIS DISTRIBUTION: Known only from Jackson and Livingston counties.

Pepoon (1927) reported this species as persisting in gardens in Naperville, but I have seen no specimens to verify this.

This attractive species is unique in the genus because of its white flowers and its bright red fruiting calyces.

Because of these lanternlike fruits, the species is a garden favorite.

The flowers bloom from July to September.

8. *Nicotiana* L. –Tobacco

Annual or perennial, usually viscid, herbs or shrubs; leaves alternate, simple, usually entire; flowers perfect, in racemes or panicles,

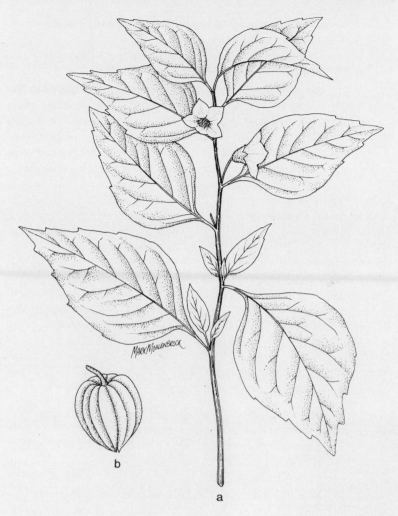

b

a

31. *Physalis alkekengi* (Chinese Lantern). *a.* Habit, × ½. *b.* Fruit, × 1.

usually opening in the evening; calyx tubular-campanulate, 5-lobed; corolla tubular, with 5 unequal lobes; stamens 5, free, attached to the corolla tube; pistil one, the ovary superior, 2-locular; fruit a 2- to 4-valved capsule with minute seeds.

Nicotiana is a genus of about one hundred species native mostly to North and South America. Most species are ill-smelling and possess poisonous narcotic substances.

KEY TO THE SPECIES OF Nicotiana IN ILLINOIS

1. Cauline leaves broadly ovate, petiolate; inflorescence paniculate; calyx lobes deltoid, about ¼ as long as the calyx tube; corolla greenish yellow, the tube up to 5 cm long _____ 1. *N. rustica*
1. Cauline leaves lanceolate to lance-ovate, sessile; inflorescence racemose; calyx lobes subulate, about as long as the calyx tube; corolla pale yellow, the tube up to 12 cm long_____ 2. *N. longiflora*

1. Nicotiana rustica L. Sp. Pl. 180. 1753. *Fig. 32.*

Annual herb; stems erect, slender, pubescent, to about 1 m tall; leaves broadly ovate, obtuse at the apex, rounded to subcordate at the base, entire, thin, pubescent, to 20 cm long, on pubescent petioles to 6 cm long; flowers in panicles, to 2.5 cm long, on pubescent pedicels to 1.5 cm long; calyx campanulate, to 8 mm long, pubescent, green, 5-lobed, the lobes deltoid, about one-fourth as long as the calyx tube; corolla tube cylindrical, with 5 shallow lobes at the widened upper end, greenish-yellow; stamens 5, included; capsule globose, glabrous, to 1 cm in diameter.

COMMON NAME: Wild Tobacco.
HABITAT: Dry waste ground.
RANGE: Native to Peru; adventive in much of North America.
ILLINOIS DISTRIBUTION: Collections have been seen from Cook, Menard, and Peoria counties.
This species contains nicotine, which is used primarily as an insecticide ingredient. It formerly was smoked by Indians.
The flowers bloom in September and October.

2. Nicotiana longiflora Cav. Descr. Pl. 106. 1802. *Fig. 33.*

Annual or perennial herb; stems erect, to 1 m tall, sparsely pubescent, rarely viscid; basal leaves in a rosette, oblanceolate to ellipticovate, pointed at the tip, tapering to the base into a winged petiole, slightly pubescent, up to 30 (to 50) cm long; cauline leaves lanceolate to lance-ovate, sessile, auriculate; inflorescence racemose, the flowers not overlapping; flowers mildly fragrant, borne on pedicels up to 20 mm long; calyx 5-lobed, the lobes subulate, about as long

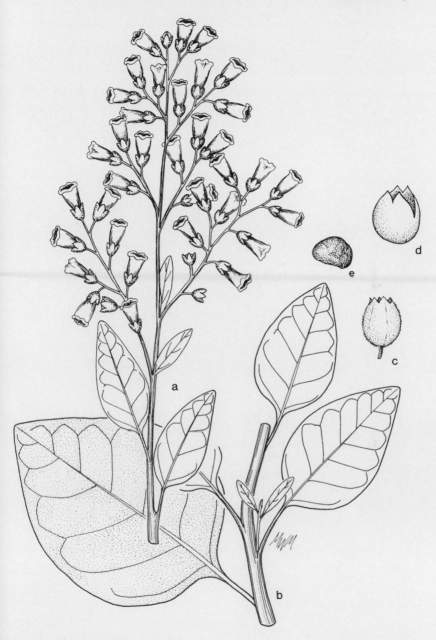

32. *Nicotiana rustica* (Wild Tobacco). *a.* Inflorescence, ×½. *b.* Leaves, ×½. *c.* Calyx, ×1¼. *d.* Fruit, ×1½. *e.* Seed, ×15.

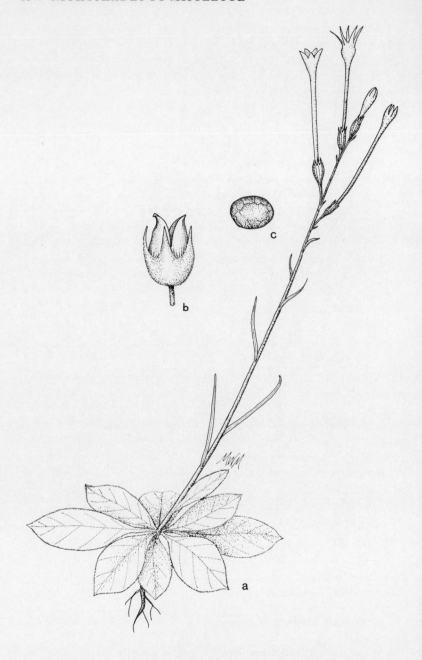

33. *Nicotiana longiflora* (Long-flowered Tobacco). *a*. Habit, × ½. *b*. Fruit, × 1½. *c*. Seed, × 12½.

as the tube, the tube 10-nerved, up to 10 mm long; corolla pale yellow, often tinged with purple, puberulent on the outside, 5-lobed, the lobes ovate, acute, the tube up to 12 cm long, up to 2.5 mm broad; capsule ovoid, 10–15 mm long, with ellipsoid seeds up to 0.5 mm long, light brown, reticulate.

COMMON NAME: Long-flowered Tobacco.

HABITAT: Waste ground (in Illinois).

RANGE: Southern South America; occasionally planted, but rarely escaped in the United States, except along the Gulf of Mexico.

ILLINOIS DISTRIBUTION: Known only from a collection by Norton in East St. Louis, St. Clair County, on July 9, 1858. The specimen is in the herbarium of the Missouri Botanical Garden.

This species differs from *N. rustica* by its racemose inflorescence, its longer yellow flowers, and its sessile cauline leaves.

It flowers during the summer.

9. *Petunia* Juss.–Petunia

Annual or perennial, usually viscid herbs; leaves alternate, entire; flowers perfect, solitary, mostly axillary; calyx deeply 5-parted; corolla funnelform or salverform, shallowly 5-lobed at the expanded apex; stamens 5, with one much smaller, free, attached to the corolla tube; pistil one, the ovary superior, 2-locular; fruit a 2-valved capsule with numerous small seeds.

Petunia is a genus of about twenty-five species, almost all native to South America.

KEY TO THE SPECIES OF Petunia IN ILLINOIS

1. Corolla white, salverform _____ 1. *P. axillaris*
1. Corolla red, violet, or purple, funnelform _____ 2
   2. Corolla 3–4 cm long _____ 2. *P. violacea*
   2. Corolla 5–9 cm long _____ 3. *P.* X *hybrida*

  1. **Petunia axillaris** (Lam.) BSP. Prel. Cat. N.Y. 38. 1898. *Fig. 34.*

Nicotiana axillaris Lam. Encycl. 4:480. 1797.

Petunia nyctaginiflora Juss. Ann. Mus. Paris 2:215, pl. 47, f. 2. 1803.

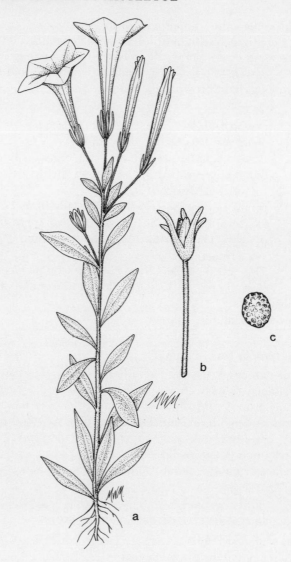

34. Petunia axillaris (White Petunia). *a.* Habit, ×½. *b.* Fruit, ×1. *c.* Seed, ×20.

Annual; stems erect, branched, stout, viscid-pubescent, to 60 cm tall; leaves ovate to obovate, obtuse to acute at the apex, cuneate to the nearly sessile base, entire, viscid-pubescent, to 9 cm long; flower solitary, axillary, on a viscid-pubescent peduncle to 9 cm long; calyx lobes 5, linear-oblong, obtuse, green, pubescent, to 1 cm long; corolla white, salverform, the tube to 3 cm long, the limb to 4.5 cm across; capsule ovoid.

COMMON NAME: White Petunia.
HABITAT: Waste ground.
RANGE: Native of South America; rarely adventive in North America.
ILLINOIS DISTRIBUTION: Scattered in Illinois.
This handsome white-flowered species blooms from July to September.

2. **Petunia violacea** Lindl. Bot. Reg. pl. 1626. 1833. *Fig. 35.*

Annual; stems erect, branched, slender, viscid-pubescent, to 40 cm tall; leaves ovate to obovate, obtuse to acute at the apex, rounded at the short-petiolate base, entire, viscid-pubescent, to 6 cm long; flower solitary, axillary, on a viscid-pubescent peduncle to 5 cm long; calyx lobes 5, linear, obtuse to subacute, green, pubescent, to 1 cm long; corolla violet or reddish, funnelform, the tube 3–4 cm long, the limb to 3 cm across; capsule ovoid.

COMMON NAME: Violet Petunia.
HABITAT: Waste ground.
RANGE: Native of South America; rarely adventive in North America.
ILLINOIS DISTRIBUTION: Occasional in various sections of Illinois.
This petunia is found more frequently as an adventive in Illinois than the other two taxa. It has smaller flowers than either *P. axillaris* or *P. X hybrida.*
The flowers bloom from June to August.

3. **Petunia X hybrida** Hort. ex. Vilm. Fl. Pl. Terre, ed. 1. 615. 1865. *Fig. 36.*

Annual; stems erect, branched, rather stout, viscid-pubescent, to 50 cm tall; leaves ovate to obovate, obtuse to acute at the apex, rounded or subcuneate at the nearly sessile or short-petiolate base, entire, viscid-pubescent, to 7.5 cm long; flower solitary, axillary, on a viscid-pubescent peduncle to 7 cm long; calyx lobes 5, linear-oblong, obtuse, green, pubescent, to 1 cm long; corolla red or

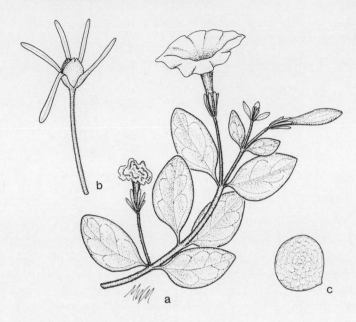

35. *Petunia violacea* (Violet Petunia). *a*. Habit, × ½. *b*. Fruit, × 1. *c*. Seed, × 30.

purple or variegated, funnelform, the tube 5–9 cm long, the limb 3–7 cm across; capsule ovoid.

COMMON NAME: Garden Petunia.

HABITAT: Waste ground.

RANGE: Horticultural plant; rarely escaped in North America.

ILLINOIS DISTRIBUTION: Scattered in Illinois.

The garden petunia is thought to be a hybrid between *P. axillaris* and *P. violacea,* although it has flowers larger than either. Numerous color variations have been developed.

The flowers bloom from June to August.

CONVOLVULACEAE–MORNING-GLORY FAMILY

Herbs, shrubs, or trees, sometimes twining, sometimes with latex; leaves alternate, simple, without stipules; inflorescence cymose, or

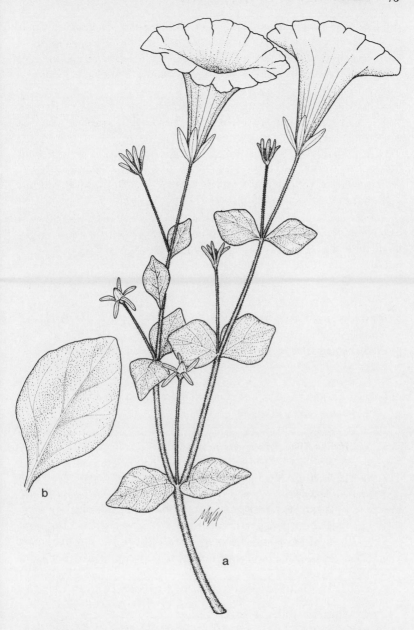

36. Petunia X *hybrida* (Garden Petunia). *a.* Habit, × ½. *b.* Leaf variation, × ½.

the flower solitary; flowers perfect, actinomorphic, bracteate; sepals 5, usually free; corolla mostly tubular, 5-lobed or entire; stamens 5, free, attached to the base of the corolla tube; disk often present; pistil 1, the ovary superior, 2-locular, with 1–2 ovules per locule on axile placentae; fruit a capsule.

This family contains about fifty genera and thirteen hundred species. The greatest concentration of species is in tropical America and tropical Asia.

The Convolvulaceae are related to the Polemoniaceae, differing by their twining habit, their latex, their free sepals, and their bicarpellate ovary with 1–2 ovules per locule. They differ from the often-included Cuscutaceae by possessing chlorophyll.

There are a number of garden ornamentals in the family. Several species of *Ipomoea*, the morning-glory, are grown for their large, handsome flowers. Another *Ipomoea*, *I. batatas* L., is the sweet potato.

KEY TO THE GENERA OF Convolvulaceae IN ILLINOIS

1. Corolla rotate, usually blue; stigmas 4 ------------------------- 6. *Evolvulus*
1. Corolla salverform, funnelform, or nearly campanulate, variously colored; stigmas 1–2 _____ 2
    2. Leaves linear, up to 3 (–6) mm broad; style deeply 2-cleft _____
    _____ 1. *Stylisma*
    2. Leaves elliptic to oblong to ovate, usually at least 6 mm broad; style undivided, although the stigmas sometimes lobed _____ 3
3. Flowers scarlet _____ 5. *Ipomoea*
3. Flowers white, blue, or purple, never scarlet _____ 4
    4. Bracts large and foliaceous, sometimes concealing the calyx ____ 5
    4. Bracts small, never exceeding the calyx, or absent_____ 6
5. Flowers in headlike clusters; calyx long-hirsute ____ 2. *Jacquemontia*
5. Flower solitary or 2–4 in a group; calyx glabrous or pubescent, not long-hirsute _____ 4. *Calystegia*
    6. Sepals up to 8 mm long; stigmas 2, linear _____ 3. *Convolvulus*
    6. Sepals 10 mm long or longer; stigmas 1 or, if 2, capitate _____
    _____ 5. *Ipomoea*

1. *Stylisma* Raf.–Stylisma

Herbaceous perennial from thickened crowns or rhizomes; stems prostrate or twining; leaves alternate, simple, entire; inflorescence cymose, axillary; flowers perfect, actinomorphic, subtended by 2 bracteoles; sepals 5, free; corolla funnelform, more or less entire; stamens 5, free; pistil 1, the ovary superior, 2-locular, surrounded by a basal disk, with ovules 2 per locule; styles deeply 2-cleft; fruit a capsule, with 1–4 seeds.

Stylisma is a genus of six species of temperate North America. Our species is sometimes placed in the genus *Breweria*, but Myint, who revised the genus *Stylisma* in 1966, has given reasons for recognizing *Stylisma* as the valid name.

Only the following taxon occurs in Illinois.

1. **Stylisma pickeringii** (Torr.) Gray var. **pattersonii** (Fern. & Schub.) Myint, Brittonia 18:114. 1966. *Fig. 37.*

Breweria pickeringii (Torr.) Gray var. *pattersoni* Fern. & Schub. Rhodora 51:42, pl. 1129. 1949.

Stylisma pattersoni (Fern & Schub.) G. N. Jones in Jones & Fuller, Vasc. Pl. Ill. 387. 1955.

Perennial herb; stems prostrate or trailing, minutely pubescent, to 2 m long; leaves linear, acute to obtuse at the apex, cuneate to the sessile or subsessile base, entire, minutely pubescent, to 7 cm long, to 3 (–6) mm wide; flowers cymose, 1–5 per inflorescence, the central ones often sessile, the lateral ones on pedicels to 2 cm long, each subtended by a pair of foliar bracts about as long as or slightly longer than the sepals; sepals 5, free, oval-lanceolate, acute or subacute, hoary-pubescent, 4–6 mm long; corolla to 1.8 cm long, white, funnelform, the margin more or less entire; stamens 5, free, partly exserted; ovary densely villous; style deeply 2-cleft, the branches 1.0–1.5 mm long, unequal; stylopodia 1–2 mm long; capsule ovoid, pubescent, to about 1 cm long, 2-seeded.

COMMON NAME: Illinois Stylisma.

HABITAT: Sandy prairies (in Illinois).

RANGE: Kansas, Oklahoma, eastern Texas; southeastern Iowa, eastern Missouri, and west-central Illinois.

ILLINOIS DISTRIBUTION: Cass, Henderson, and Mason counties. The type collection for this taxon was made by H. N. Patterson from a sandy prairie four miles north of Oquawka, Henderson County, on August 11, 1873. It was subsequently refound at the same locality by V. H. Chase in 1934. Since that time, it has been discovered in Cass and Mason counties. The Illinois

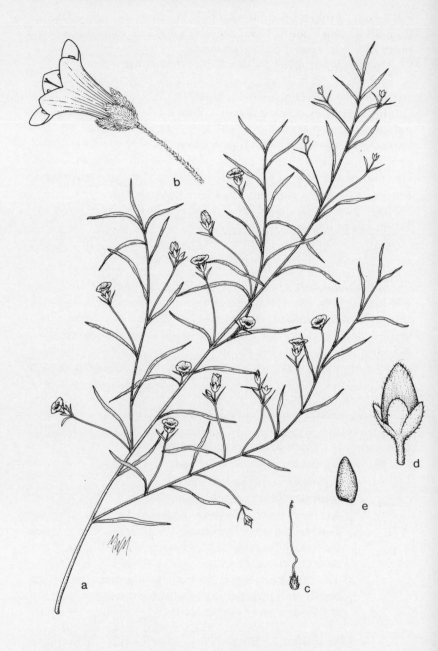

37. *Stylisma pickeringii* var. *pattersonii* (Illinois Stylisma). *a.* Habit, × ½. *b.* Flower, × 2½. *c.* Pistil, × 2. *d.* Fruit, × 2½. *e.* Seed, × 2½.

collections, a collection from Allenton, Missouri, and one collection from Muscatine County, Iowa, represent disjunct populations in the range of S. *pickeringii* var. *pattersonii*.

The handsome white flowers bloom during August.

2. *Jacquemontia* Choisy–Jacquemontia

Trailing annual or perennial herbs; leaves alternate, entire or undulate; inflorescence cymose, axillary, 1- to several-flowered, subtended by several densely arranged bracts; calyx deeply 5-parted, or the sepals free; corolla campanulate, 5-lobed; stamens 5, free, attached to the base of the corolla; pistil 1, the ovary superior, 2-locular, the stigmas ellipsoid; fruit a 2-locular capsule.

Jacquemontia is a genus of few species in the southern United States and in tropical America.

Only the following adventive species has been found in Illinois.

1. Jacquemontia tamnifolia (L.) Griseb. Fl. Brit. W. Ind. 474. 1864. *Fig. 38.*
Ipomoea tamnifolia L. Sp. Pl. 161. 1753.

Trailing annual; stems tawny-pubescent, branched; leaves ovate to oblong-ovate, acuminate at the tip, rounded or cordate at the base, entire to undulate, pubescent, up to 12 cm long; flowers in axillary, capitate cymes up to 3 cm broad, borne on peduncles at least as long as the subtending leaves, with several dense bracts; sepals 5, rusty-pubescent, nearly equal, linear, up to 9 mm long; corolla campanulate, blue, 5-lobed, up to 3 cm across; capsule subglobose, up to 6 mm in diameter, with brownish black seeds.

COMMON NAME: Jacquemontia.

HABITAT: Along a railroad (in Illinois).

RANGE: Virginia to Arkansas, south to Mississippi and Florida; adventive in Illinois.

ILLINOIS DISTRIBUTION: Known only from along a railroad in Grundy County.

The capitate-cymose inflorescence and the rusty hairy sepals readily distinguish this rare adventive.

It flowers during the summer.

3. *Convolvulus* L. –Bindweed
Perennial herbs; stems trailing or twining; leaves alternate, usually entire, sometimes sagittate or hastate; inflorescence cymose, axil-

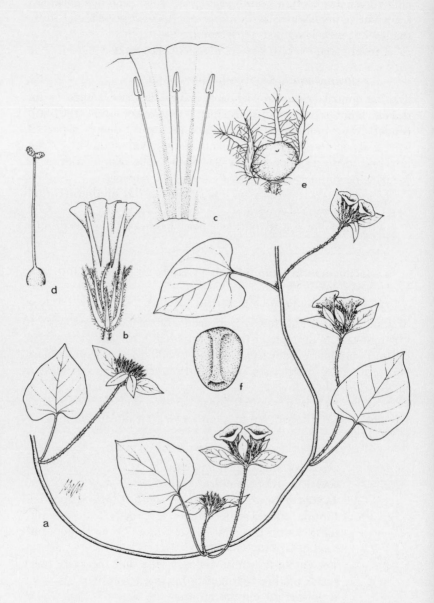

38. *Jacquemontia tamnifolia* (Jacquemontia). *a.* Habit, × ½. *b.* Flower, with bracts, × 2½. *c.* Part of corolla (opened out), × 5. *d.* Pistil, × 5. *e.* Fruit with some bracts (sepals removed), × 2½. *f.* Seed, × 7½.

lary, 1- to 4-flowered; flowers perfect, actinomorphic, without leafy bracts; calyx deeply 5-parted, or the sepals free; corolla funnelform, shallowly 5-lobed; stamens 5, free, attached to the base of the corolla; pistil 1, the ovary superior, 2-locular, the stigmas filiform; fruit a 2-locular capsule.

I am following Lewis and Oliver (1965) and Brummitt (1965) in dividing the traditional genus *Convolvulus* into *Convolvulus*, with the calyx not subtended by a pair of leafy bracts, and *Calystegia*, with the calyx bibracteate.

KEY TO THE SPECIES OF Convolvulus IN ILLINOIS

1. Stems glabrous; leaves linear-oblong to ovate, glabrous or nearly so; sepals to 3 mm long; corolla to 2.5 cm broad _____ 1. *C. arvensis*
1. Stems cinereous-pubescent; leaves oblong to elliptic, cinereous-pubescent; sepals to 8 mm long; corolla to 1.2 cm broad_____
_____ 2. *C. incanus*

1. **Convolvulus arvensis** L. Sp. Pl. 153. 1753. *Fig. 39.*

Convolvulus ambigens House, Bull. Torrey Club 32:139. 1905.

Perennial herb from a deep, slender rootstock; stems trailing, branched, glabrous or nearly so, to 1 m long; leaves linear-oblong to ovate, obtuse to subacute at the apex, cordate, hastate, or sagittate at the base, entire, glabrous or nearly so, to 5 cm long; flowers 1–2 from the axils, without leafy bracts, but sometimes with minute bracts low on the pedicel; sepals 5, more or less free, oblong, obtuse, green, to 3 mm long, glabrous; corolla funnelform, white or pinkish, to 2 cm long, the limb to 2.5 cm broad, shallowly 5-lobed; stigmas filiform; capsule glabrous, with up to 4 seeds.

COMMON NAME: Field Bindweed.

HABITAT: Waste ground.

RANGE: Native of Europe; naturalized throughout North America.

ILLINOIS DISTRIBUTION: Occasional to common throughout the state.

The field bindweed is a common weed of virtually every kind of waste area. It can be a troublesome weed in cultivated fields.

Considerable variation occurs in the leaves, with both hastate and sagittate forms known.

39. *Convolvulus arvensis* (Field Bindweed). *a*. Habit, × ½. *b*. Leaf variations, × ½.
c. Variation, × 5. *d*. Flower (opened out and partially cut away), × 2. *e*. Stamen, × 5.
f. Fruit, × 2½. *g*. Seed, × 2½.

The glabrous stems and leaves and larger corolla distinguish this species from *C. incanus.*

The white or pinkish flowers bloom from May to September.

2. Convolvulus incanus Vahl, Sym. Bot. 3:23. 1794. *Fig. 40.*

Trailing perennial; stems cinereous-pubescent, branched from the base; leaves oblong to elliptic, more or less rounded at the tip, tapering to a short petiole, cinereous-pubescent, up to 2.5 cm long; flowers 1–several from the axils of the leaves; sepals 5, more or less free, green, up to 8 mm long; corolla white, up to 1.2 cm across, with 5 acute lobes; capsule ovoid, up to 5 mm long, splitting at maturity into several valves.

COMMON NAME: Ashy Bindweed.

HABITAT: Dry soil.

RANGE: Nebraska to Colorado, south to Arizona and Arkansas; apparently adventive in Illinois.

ILLINOIS DISTRIBUTION: St. Clair Co.: along railroad north of East St. Louis, *N. M. Glatfelter s.n.*

This species, apparently adventive from the West, differs from *C. arvensis* by its ashy pubescence and smaller flowers.

The specimen, deposited in the herbarium of the Missouri Botanical Garden, has been annotated by Dr. H. Hallier.

Kartesz and Kartesz (1980) consider *C. incanus* to be a synonym of *C. arvensis.*

The flowers bloom during the summer.

4. *Calystegia* R. Br.–Bindweed

Perennial herbs; stems trailing, twining, or erect; leaves alternate, entire or basally lobed; inflorescence cymose, axillary, 1- to 4-flowered; flowers perfect, actinomorphic, subtended by a pair of foliar bracts; calyx deeply 5-parted, or the sepals free; corolla funnelform, shallowly 5-lobed or entire; stamens 5, attached to the base of the corolla; pistil 1, the ovary superior, 2-locular, the stigmas ovoid to ellipsoid; fruit a 2-locular capsule.

Calystegia is distinguished from *Convolvulus* by the pair of foliar bracts subtending each flower and by the ovoid or ellipsoid stigmas.

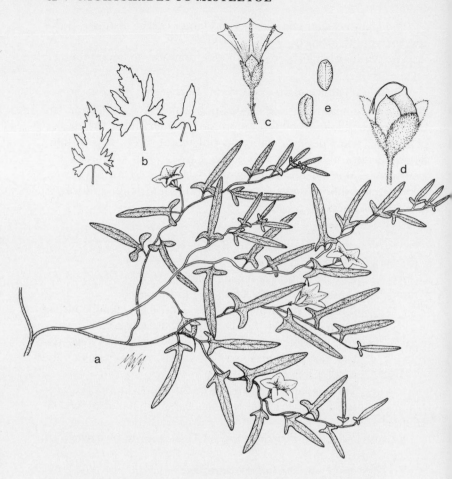

40. *Convolvulus incanus* (Ashy Bindweed). *a.* Habit, × ½. *b.* Leaf variations, × ½. *c.* Flower, × 1. *d.* Fruit, × 1½. *e.* Seeds, × 1½.

KEY TO THE SPECIES OF Calystegia IN ILLINOIS

1. Flowers double _____ 1. *C. pubescens*
1. Flowers single _____ 2
    2. Plants upright, either erect or ascending _____ 2. *C. spithamaea*
    2. Plants trailing or climbing _____ 3
3. Peduncles, or most of them, longer than the petioles; bracteoles usually not overlapping, acute, not saccate_____ 3. *C. sepium*

3. Peduncles, or most of them, shorter than the petioles; bracteoles overlapping, obtuse, saccate _____ 4.*C. silvatica*

1. **Calystegia pubescens** Lindl. Journ. Hort. Soc. 1:70. 1846.
   Fig. 41.
   Convolvulus pellitus Ledeb. f. *anestius* Fern. Rhodora 51:73.
   1949.

Perennial herb from a deep, slender rootstock; stems twining,
branched, softly pubescent, to 7 m long; leaves lanceolate, obtuse
to acute at the apex, hastate or narrowly sagittate at the base, to 10
cm long, softly pubescent, the petioles to 3 cm long; flowers solitary, axillary, subtended by a pair of ovate, foliaceous bracts to 1.5
cm long; sepals 5, free, green, oblong, obtuse; corolla bright pink,
to 5 cm long, to 4.5 cm across, composed of numerous segments;
stamens absent; pistils absent.

COMMON NAME: California Rose.
HABITAT: Waste ground.
RANGE: Native of Asia; escaped from cultivation in the
eastern half of the United States.
ILLINOIS DISTRIBUTION: Known only from Champaign, Du Page, Wabash, and Washington counties.
This handsome double-flowered form has flowers that
persist for several days. It is sterile.
This species has been confused with *Convolvulus ja-
ponicus* Thunb., another cultivated double-flowered
plant.

The flowers of *Calystegia pubescens* bloom from May to September.

2. **Calystegia spithamaea** (L.) Pursh, Fl. Am. Sept. 143. 1814.
   Fig. 42.
   Convolvulus spithamaeus L. Sp. Pl. 158. 1753.

Perennial herb; stems ascending to erect, rarely trailing, branched
from near the base, glabrous or pubescent, to 45 cm tall; leaves
narrowly oblong to oval, obtuse at the apex, rounded at the sessile
or short-petiolate base, entire, glabrous or pubescent, to 8 cm long,
the lower leaves smaller than the upper ones; flowers 1–4 from the
lowermost axils, on peduncles longer than the adjacent leaves, subtended by a pair of oval, foliaceous bracts longer than the calyx;

41. *Calystegia pubescens* (California Rose). *a.* Habit, × ½. *b.* Flower, × 1.

sepals 5, free or nearly so, green, oblong, to 1 cm long; corolla white, to 7 cm long, to nearly as broad, shallowly 5-lobed; stamens 5, included to barely exserted; stigmas oblongoid; capsule mostly 4-seeded.

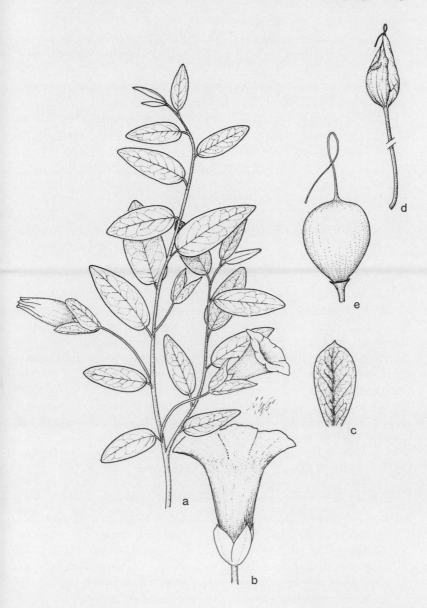

42. *Calystegia spithamaea* (Dwarf Bindweed). *a.* Habit, × ½. *b.* Flower (bracts removed), × 1. *c.* Bract, × 1. *d.* Fruit with enclosing bracts, × 1. *e.* Fruit (bracts removed), × 1.

COMMON NAME: Dwarf Bindweed.

HABITAT: Open woods, prairies, sandy soil.

RANGE: Quebec to Minnesota, south to Missouri and Virginia.

ILLINOIS DISTRIBUTION: Occasional in the northern four-fifths of the state; also Hardin County.

This is the only species of *Calystegia* in Illinois that grows upright. The large white flowers, borne by a plant of such short stature, are striking.

The lowermost leaves are much smaller than the upper ones.

The flowers are borne from May to July.

3. **Calystegia sepium** (L.) R. Br. Prodr. Fl. Nov. Holl. 1:483. 1810.

Convolvulus sepium L. Sp. Pl. 153. 1753.

Perennial herb; stems twining or trailing, branched, glabrous or pubescent, to 4 m long; leaves oblong to ovate to deltoid, obtuse to acute to acuminate at the apex, cordate to hastate to sagittate at the base, entire or sparsely dentate, glabrous or pubescent, to 10 cm long, on glabrous to pubescent petioles to 5 cm long; flowers 1–2 per axil, on peduncles longer than the petioles, subtended by a pair of ovate, cordate, foliaceous bracts to 3.5 cm long; sepals 5, free or nearly so, green, glabrous to pubescent, concealed by the bracts; corolla white to pinkish, to 8 cm long, nearly as broad, shallowly 5-lobed; stamens 5, included to barely exserted; stigmas oblongoid; capsule up to 4-seeded.

Typical var. *sepium* is native to Europe and Asia. The Illinois plants can be separated into several subspecies.

KEY TO SUBSPECIES OF C. sepium IN ILLINOIS

1. Bracteoles not clearly distinct from the sepals, forming a continuous spiral with the sepals and gradually merging with them; sinus of leaves strongly occluded _____ 3d. *C. sepium* ssp. *erratica*
1. Bracteoles clearly differentiated from the sepals, never merging with them; sinus of leaves open or rarely only slightly occluded _____ 2
   2. Plants glabrous or sparsely pubescent; leaves mostly hastate ____ 3
   2. Plants densely soft-pubescent; leaves sagittate_____
   _____ 3. *C. sepium* var. *repens*
3. Leaves with V-shaped sinus; corolla sometimes pink _____
   _____ 3a. *C. sepium* ssp. *americana*

3. Leaves with U-shaped sinus; corolla white (rarely suffused with pale
   rose)_____ 3b. *C. sepium* ssp. *angulata*

   3a. Calystegia sepium (L.) R. Br. ssp. **americana** (Sims) Brum-
   mitt, Ann. Mo. Bot. Gard. 52:216. 1965. *Fig. 43a–e.*
   Convolvulus sepium L. var. *americanus* Sims, Bot. Mag. 19:pl.
   732. 1804.
   Convolvulus americanus (Sims) Greene, Pittonia 3:328. 1898.
   Convolvulus sepium L. var. *communis* Tryon, Rhodora 41:419.
   1939.
   Calystegia sepium (L.) R. Br. var. *americana* (Sims) Kitagawa,
   Rep. Inst. Sci. Res. Manchoukuo 3 App. 1:365. 1939.

Plants glabrous or sparsely pubescent; leaves mostly hastate, with a
V-shaped sinus; bracteoles clearly differentiated from the sepals.

COMMON NAME: American Bindweed.
HABITAT: Moist soil, fields, waste areas.
RANGE: Newfoundland to British Columbia, south to
Oregon, New Mexico, and Florida.
ILLINOIS DISTRIBUTION: Common throughout the
state; probably in every county.
This is the most common of the subspecies of *C. sepium*
in Illinois. It occurs in a wide variety of habitats, from
waste ground to calcareous meadows to prairies.
Subspecies *americana* differs from ssp. *angulata* by the
V-shaped sinus between the basal lobes of the leaf and
by its sometimes pink flowers. The two subspecies show some intergradation, although most specimens can be assigned readily to
one subspecies or the other.
The flowers appear from June to August.

   3b. Calystegia sepium (L.) R. Br. ssp. **angulata** Brummitt, Kew
   Bull. 35(2):328. 1980. *Fig. 43f.*
Plants glabrous; leaves mostly hastate, with a U-shaped sinus; corolla white, rarely tinged with pale rose; bracteoles clearly differentiated from the sepals.

43. *Calystegia sepium* ssp. *americana* (American Bindweed). *a.* Habit, × ½. *b.* Flowers with bracteoles removed, showing calyx, × ½. *c.* Bracteole, × 1. *d.* Corolla (partly cut away), × ½. *e.* Pistil with annular disc at base, × 2½. ssp. *angulata* (Bindweed). *f.* Leaf, × ½. var. *repens* (Trailing Bindweed). *g.* Leaf, × ½. ssp. *erratica* (Erratic Bindweed). *h.* Leaf, × ½. *i.* Flower, × ½. *j.* Progression of bracteoles (left) to sepals (right), × ½.

COMMON NAME: Bindweed.

HABITAT: Roadsides, fields.

RANGE: British Columbia to California, east through the Rocky Mountains and Great Plains to New England.

ILLINOIS DISTRIBUTION: Known from Cook and Woodford counties.

Until Brummitt (1980) described this subspecies, specimens referable to it had been called ssp. *americana*. It is more of a western subspecies, differentiated from ssp. *angulata* by its U-shaped sinus and usually white flowers. *Jones 14316*, however, from Woodford County has pink-tinged flowers.

The flowers bloom from June through August.

3c. Calystegia sepium (L.) R. Br. var. **repens** (L.) Gray Man. Bot. N. U.S. 348. 1898. *Fig. 43g.*

Convolvulus repens L. Sp. Pl. 158. 1753.

Plants densely soft-pubescent; leaves mostly sagittate.

COMMON NAME: Trailing Bindweed.

HABITAT: Along a railroad.

RANGE: Ohio to Wisconsin, south to Texas and Florida; New Brunswick.

ILLINOIS DISTRIBUTION: Lake Co.: Diamond Lake, July 1, 1907, *F. C. Gates 1716.*

This softly pubescent variety, whose natural range should include Illinois, is known in this state only from along a railroad.

The flowers bloom during July.

3d. Calystegia sepium (L.) R. Br. ssp. **erratica** Brummitt, Kew Bull. 35(2):330. 1980. *Fig. 43h, i, j.*

Plants glabrous or often pubescent; lobes of leaves truncate, rarely spreading; sinus of leaves strongly occluded; bracteoles not clearly differentiated from the sepals but forming a continuous spiral with the sepals and gradually merging with them; bracteoles green, acute to subobtuse, carinate, 1.6–2.6 cm long, 1.0–2.4 cm broad; corolla rose, 4.3–6.0 cm long; stamens 2.5–3.0 cm long.

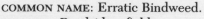

COMMON NAME: Erratic Bindweed.

HABITAT: Roadsides, fields.

RANGE: Ontario, New York, New Jersey, Pennsylvania, Michigan, Indiana, Illinois; adventive in Oregon.

ILLINOIS DISTRIBUTION: Known from Richland County. The type for this subspecies was collected ten miles northeast of Olney in Richland County, Illinois, on May 25, 1925, by Robert Ridgway.

Subspecies *erratica* differs from all other subspecies of *C. sepium* by its bracteoles that form a continuous spiral and merge imperceptibly into the sepals so that each flower appears to be subtended by 3–4 bracteoles, rather than 2.

The flowers bloom from May through August.

4. **Calystegia silvatica** (Kit.) Griseb. ssp. **fraterniflorus** (Mack. & Bush) Brummitt, Kew Bull. 35(2):332. 1980. *Fig. 44.*

Convolvulus sepium L. var. *fraterniflorus* Mack. & Bush, Man. Fl. Jackson Co., Mo. 153. 1902.

Convolvulus fraterniflorus (Mack. & Bush) Mack. & Bush, Ann. Rep. Mo. Bot. Gard. 16:104. 1905.

Calystegia fraterniflora (Mack. & Bush) Brummitt, Ann. Mo. Bot. Gard. 52:216. 1965.

Calystegia sepium (L.) R. Br. var. *fraterniflora* (Mack. & Bush) Shinners, Sida 3:282. 1968.

Perennial herb; stems twining or trailing, branched, glabrous or pubescent, to 3 m long; leaves oblong to ovate to deltoid, obtuse to acute at the apex, more or less hastate at the base, usually entire, glabrous or pubescent, to 12 cm long, on petioles up to 5 cm long; flowers 1–2 per axil, on peduncles often shorter than the petioles, subtended by saccate, overlapping, obtuse bracteoles; sepals 5, free or nearly so, green, glabrous to pubescent; corolla white to pinkish, to 10 cm long, nearly as broad, shallowly 5-lobed; stamens 5, usually included; capsule up to 4-seeded.

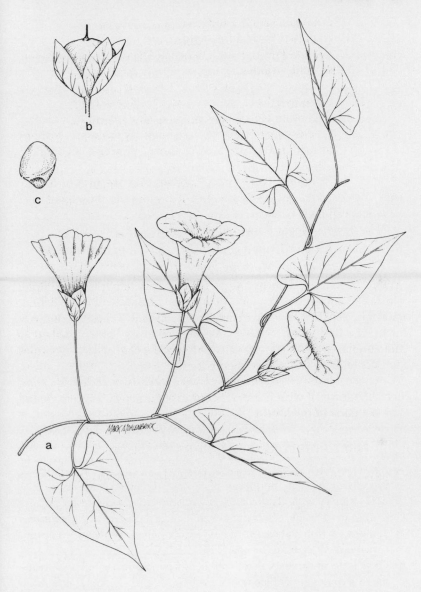

44. Calystegia silvatica ssp. *fraterniflorus* (Bindweed). *a.* Habit, × ½ . *b.* Fruit, × 1.
c. Seed, × 5.

COMMON NAME: Bindweed.
HABITAT: Roadsides, fields.
RANGE: Pennsylvania to North Dakota, south to Arkansas and Virginia.
ILLINOIS DISTRIBUTION: Occasional in the southern three-fourths of the state; also DeKalb County.
This taxon usually has been considered either a distinct species or a variety of *C. sepium*, differing primarily by its peduncles, or most of them, shorter than its petioles.

Brummitt (1980) now believes that because of its saccate, overlapping, obtuse bracteoles, it is more nearly related to *C. silvatica*, calling it *C. silvatica* ssp. *fraterniflorus*.

The flowers bloom from June to August.

5. *Ipomoea* L. –Morning-glory

Annual or perennial herbs; stems trailing or twining; leaves simple, alternate; flowers solitary or in cymes or racemes, perfect, actinomorphic; sepals 5, free or nearly so; corolla salverform to campanulate, the limb more or less 5-lobed; stamens 5, free, attached to the corolla; pistil 1, the ovary superior, 2-locular, with 2–3 ovules per locule; fruit a 2- to 4-seeded capsule.

As considered here, the genus *Quamoclit* is included in *Ipomoea*. About five hundred species comprise *Ipomoea*. They are found in most parts of the World.

KEY TO THE SPECIES OF *Ipomoea* IN ILLINOIS

1. Flowers scarlet; stamens and style exserted _____ 1. *I. coccinea*
1. Flowers white, pink, purple, or blue; stamens and style included __ 2
    2. Perennials with glabrous or puberulent stems; sepals glabrous, obtuse to subacute; seeds pubescent _____ 2. *I. pandurata*
    2. Annuals with pubescent stems; sepals pubescent, acute to acuminate; seeds glabrous or nearly so _____ 3
3. Calyx lobes linear-lanceolate, with long-tapering tips; corolla essentially 3.0–4.5 cm long, sky blue (when fresh) _____ 3. *I. hederacea*
3. Calyx lobes oblong to lanceolate, acute to short-acuminate; corolla either less than 3 cm long or more than 4.5 cm long, not sky blue (when fresh) _____ 4
    4. Corolla less than 3 cm long, essentially white; ovary and capsule 2-locular _____ 4. *I. lacunosa*

4. Corolla more than 4.5 cm long, usually not white; ovary and capsule 3-locular ------------------------------- 5. *I. purpurea*

1. **Ipomoea coccinea** L. Sp. Pl. 160. 1753. *Fig. 45.*
   Quamoclit coccinea (L.) Moench, Meth. 453. 1794.

Annual herb; stems twining, slender, glabrous to puberulent, to 6 m long; leaves ovate, acuminate at the apex, cordate at the base, entire or angulate, glabrous or puberulent, to 15 cm long, the petioles to 12 cm long; flowers 2 or more on axillary peduncles; sepals 5, free, oblong, subulate-tipped, green, usually glabrous, to 4 (–7) mm long; corolla salverform, scarlet, to 3 cm long, 5-lobed; stamens 5, barely exserted; capsule globose, 4-locular, 4-valved, to 8 mm in diameter, with 4 seeds.

COMMON NAME: Red Morning-glory.
HABITAT: Fields, roadsides.
RANGE: Native of tropical America; naturalized in most of the United States.
ILLINOIS DISTRIBUTION: Occasional in the southern half of the state; rare elsewhere.

This is a handsome species whose scarlet flowers bloom from July to October. It is closely related to the ornamental Cypress Vine (*I. quamoclit* L.), a species with pinnately parted leaves.

Some botanists place this species and the cypress vine in the genus *Quamoclit*.

A collection by Ridgway (#1429) from Richland County is unusual in having the calyx up to 7 mm long.

2. **Ipomoea pandurata** (L.) Meyer, Prim. Fl. Esseq. 100. 1818. *Fig. 46.*
   Convolvulus panduratus L. Sp. Pl. 153. 1753.

Perennial herb from fleshy roots; stems trailing or climbing, branched, glabrous or puberulent, to 5 m long; leaves ovate, acute at the apex, cordate at the base, entire or angulate or rarely 3-lobed, glabrous or puberulent, to 15 cm long, on petioles to 10 cm long; flowers 1–5 per peduncle; sepals 5, free, oblong, obtuse to subacute, glabrous, green, to 2 cm long; corolla funnelform, to 8 cm long, white with purple stripes in the throat; ovary 2-locular; stigmas 2; capsule ovoid, glabrous, with 2–4 seeds densely woolly on the margins.

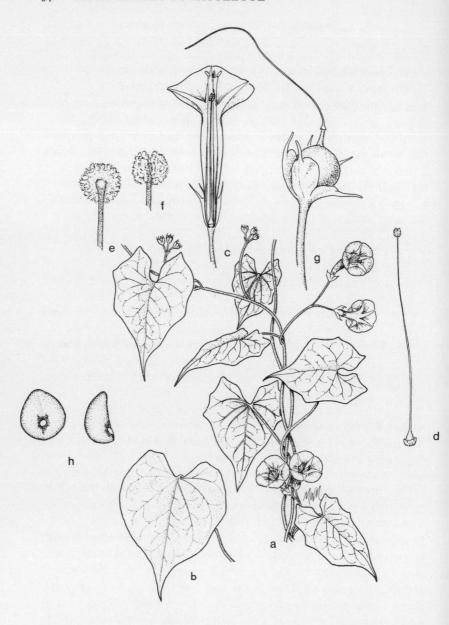

45. *Ipomoea coccinea* (Red Morning-glory). *a.* Habit, × ½. *b.* Leaf variation, × ½.
c. Flower, × 1½. *d.* Pistil, × 2½. *e.* Close-up of stigma showing central attachment,
× 20. *f.* Close-up of stigma from side, × 20. *g.* Fruit, × 2½. *h.* Seeds, × 3½.

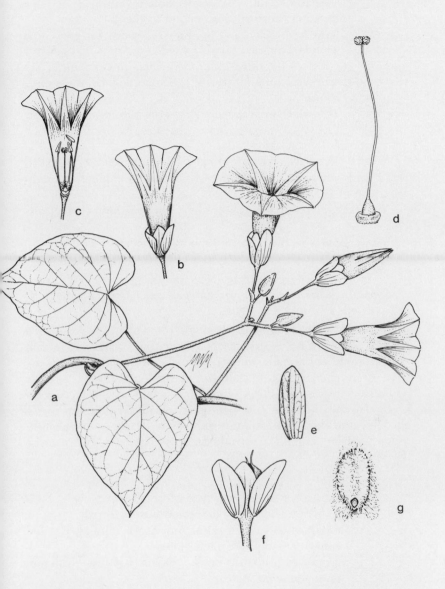

46. Ipomoea pandurata (Wild Sweet Potato Vine). *a.* Habit, ×½. *b.* Flower, with bracts removed, ×½. *c.* Flower (partially cut away), ×½. *d.* Pistil, ×2. *e.* Bract, ×1. *f.* Fruit (mostly hidden by calyx and bracts), ×3. *g.* Seed, ×2½.

COMMON NAME: Wild Sweet Potato Vine.

HABITAT: Disturbed areas and thickets.

RANGE: Connecticut to Kansas, south to Texas and Florida; Ontario.

ILLINOIS DISTRIBUTION: Occasional to common in the southern three-fourths of the state, less common elsewhere.

The wild sweet potato vine grows from a fleshy root that sometimes reaches a great size, reportedly as much as two feet long.

This species can become a troublesome weed in cultivated fields where its underground structures are difficult to eradicate.

The large, glabrous sepals serve as a distinguishing characteristic.

The large flowers bloom from June to October.

3. **Ipomoea hederacea** Jacq. Icon. Rar. pl. 36. 1781. *Fig. 47.*
Ipomoea hederacea Jacq. var. *integriuscula* Gray, Syn. Fl. N. Am., ed. 2, 2:pt. 1:433. 1886.

Annual herb; stems twining or climbing, retrorsely pubescent, to 2 m long; leaves ovate or suborbicular in outline, acuminate at the apex, cordate at the base, deeply 3-lobed, rarely unlobed, pubescent, to 12 cm long, on petioles to 15 cm long; flowers 1–3 per peduncle; sepals 5, free, linear-lanceolate, with recurved-acuminate tips, hirsute, to 2.5 cm long; corolla funnelform, to 4.5 cm long, the tube whitish, the limb sky-blue fading to purplish; ovary 3-locular; stigmas 3; capsule depressed-globose, to 1.5 cm in diameter.

COMMON NAME: Ivy-leaved Morning-glory.

HABITAT: Cultivated fields and waste ground.

RANGE: Native of tropical America; naturalized in the eastern half of the United States.

ILLINOIS DISTRIBUTION: Occasional to common throughout the state.

Most specimens of *I. hederacea* have deeply 3-lobed leaves, but occasionally entire-leaved plants are seen. These are sometimes designated var. *integriuscula*.

The most common habitat for this species in Illinois is in cultivated fields.

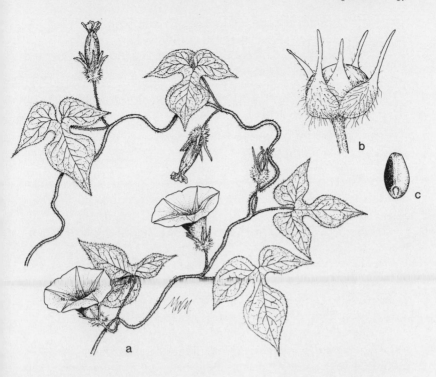

47. Ipomoea hederacea (Ivy-leaved Morning-glory). *a*. Habit, ×½. *b*. Fruit, ×1½. *c*. Seed, ×2½.

The hirsute, long-acuminate sepals are an easy identifying character.

The first reports of this species from Illinois were called *I. nil* (L.) Roth, but this binomial was erroneously applied.

The beautiful sky-blue flowers quickly fade to purplish. They bloom from June to October.

4. Ipomoea lacunosa L. Sp. Pl. 161. 1753. *Fig. 48.*

Ipomoea lacunosa L. f. *purpurata* Fern. Rhodora 40:454. 1938.

Annual herb; stems twining, glabrous or sparsely pubescent, to 3 m long; leaves ovate, acuminate at the apex, cordate at the base, entire, angled, or 3-lobed, glabrous to sparsely pubescent, to 10 cm long, on petioles usually as long as the blades; flowers 1–3 per peduncle; sepals 5, free, oblong to lanceolate, acute, green, pubescent, to 1 cm long; corolla funnelform, white or purplish, to 3 cm

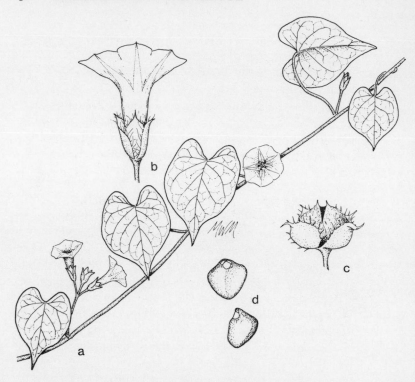

48. *Ipomoea lacunosa* (Small White Morning-glory). *a.* Habit, × ½. *b.* Flower, × 1½. *c.* Fruit, × 1½. *d.* Seeds, × 2½.

long; ovary 2-locular; stigmas 2; capsule globose, to 1 cm in diameter, usually pubescent, the seeds glabrous.

COMMON NAME: Small White Morning-glory.

HABITAT: Fields and near streams.

RANGE: Pennsylvania to Kansas, south to Texas and Georgia.

ILLINOIS DISTRIBUTION: Occasional in the southern four-fifths of Illinois, apparently absent elsewhere.

This species has relatively small white flowers, although on rare occasions purple-flowered specimens are encountered. These have sometimes been designated f. *purpurata.* There is also variation in degree of pubescence of the leaves and stems.

The flowers are borne from July to October.

5. **Ipomoea purpurea** (L.) Lam. Tabl. Encycl. 1:466. 1791. *Fig. 49.*

Convolvulus purpureus L. Sp. Pl. ed. 2, 219. 1762.

Annual herb; stems twining or trailing, retrorsely hairy to nearly glabrous, to 3 m long; leaves ovate, acute to acuminate at the apex, cordate at the base, entire, pubescent, to 10 cm long, on petioles often as long as the blades; flowers 1–5 per peduncle; sepals 5, free, oblong to lanceolate, acute, green, pubescent, to 1.8 cm long; corolla funnelform, blue, purple, red, or variegated, to 7 cm long; ovary 3-locular; stigmas 3; capsule globose, to 1 cm in diameter, glabrous, the seeds pubescent.

COMMON NAME: Common Morning-glory.

HABITAT: Disturbed areas.

RANGE: Native of tropical America; naturalized throughout North America.

ILLINOIS DISTRIBUTION: Occasional throughout the state.

The large purplish flower and 3-locular ovary distinguish this *Ipomoea* from any others in the state.

It occurs in a wide variety of weedy habitats, and is particularly common along railroads.

The flowers bloom from July to October.

6. *Evolvulus* L. –Ascending Morning-glory

Herbaceous perennial, from thick roots; stems never twining; leaves alternate, simple, entire; flower solitary or few in the axils of the leaves, perfect, actinomorphic, usually without bracts, sepals 5, free; corolla rotate; stamens 5, attached near the top of the short corolla tube; pistil 1, the ovary superior, 2-locular, with ovules 2 per locule; styles 2, each divided to form 4 stigmas; fruit a capsule, with 1–4 seeds.

Although there are approximately one hundred species of this genus in subtropical and tropical America, only the following adventive species occurs in Illinois.

1. **Evolvulus nuttallianus** Schultes, Syst. Veg. 6:198. 1820. *Fig. 50.*

Evolvulus pilosus Nutt. Gen. Am. 1:174. 1818, *in synon.*

Perennial herb from a deep root; stems ascending to more or less erect, up to 20 cm tall, densely cinereous-pubescent; leaves lanceolate to narrowly oblong, acute to subacute at the apex, tapering to

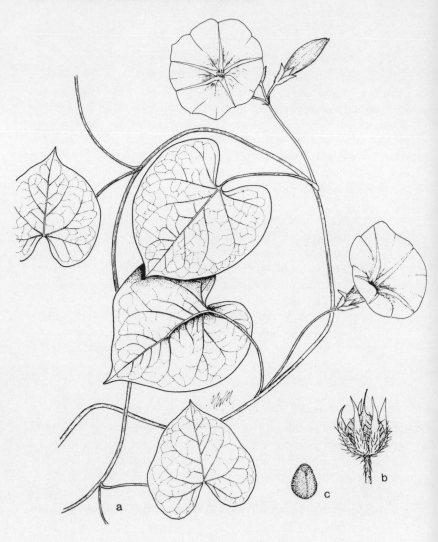

49. *Ipomoea purpurea* (Common Morning-glory). *a.* Habit, × ½. *b.* Fruit, × 1. *c.* Seed, × 2.

the base, up to 2 cm long, up to 5 mm broad, densely cinereous-pubescent, sessile or on very short petioles; flower solitary in the axils of the leaves, sessile or on short, pubescent pedicels; sepals 5, narrowly lanceolate, green, villous, up to 3 mm long; corolla rotate, blue, up to 1 cm across, the margins more or less entire; stamens 5, attached to the upper part of the corolla; stigmas 4; capsule ovoid.

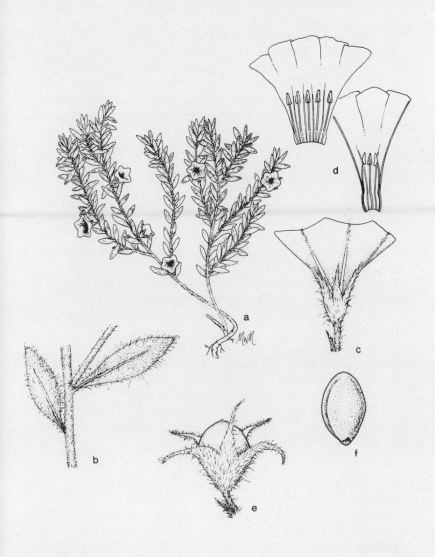

50. *Evolvulus nuttallianus* (Ascending Morning-glory). *a.* Habit, ×½. *b.* Leaves, ×2½. *c.* Flower, ×5. *d.* Flowers (partly cut away), ×5. *e.* Fruit, ×5. *f.* Seed, ×7½.

COMMON NAME: Ascending Morning-glory.

HABITAT: Disturbed soil.

RANGE: North Dakota to Montana, south to Arizona and Missouri; adventive in Illinois.

ILLINOIS DISTRIBUTION: Kane Co.: west of Montgomery, July 4, 1976, *R. Young s.n.*

This adventive from the western United States is associated with *Callirhoe involucrata* and *Penstemon cobaea* at its Illinois location.

Although *Evolvulus pilosus* Nutt. is sometimes used for this species, it is apparently an invalid binomial.

It flowers during the summer.

CUSCUTACEAE–DODDER FAMILY

Twining, annual parasitic herbs, without chlorophyll, attached to hosts by means of haustoria; stems threadlike, orange or yellow; leaves scalelike, much reduced, without chlorophyll; flowers in cymes, sometimes densely crowded and resembling a head or spike, perfect, actinomorphic; calyx 4- or 5-parted; corolla 4- or 5-parted, white, with a diversity of fringed or cleft scales on the tube beneath each stamen; stamens 4–5, alternating with the lobes of the corolla; ovary superior, 2-locular, with 2 free or united styles; fruit a capsule.

Only the genus *Cuscuta*, with 150 species found worldwide, comprises the family. All species are noxious and harmful weeds that severely parasitize their hosts.

Although *Cuscuta* is traditionally placed in the Convolvulaceae, I am following Cronquist (1981) in recognizing it in its own family because of its absence of chlorophyll and its anatomical and chemical differences.

1. *Cuscuta* L. –Dodder

Parasitic annual herbs; stems twining, with minute suckers, orange or yellow or red; leaves reduced to alternate scales, or absent; flowers in cymes, perfect, actinomorphic; calyx 4- to 5-parted, or the sepals free; corolla campanulate to urceolate, 4- to 5-parted; stamens 4–5, free, attached to the corolla; pistil 1, the ovary superior, 2-locular, with 2 ovules per locule; styles 2, usually free; capsule circumscissile, 1- to 4-seeded, the seeds glabrous.

Species differentiation is often difficult, and use of a hand lens is usually necessary for identification. Often there are petaloid scales between the lobes of the corolla that need to be examined.

Various species have been found on a wide variety of herbs and shrubs. Usually more than one kind of species can serve as host to a species of *Cuscuta*.

The seeds of *Cuscuta* germinate in the soil. Upon emerging from the ground, the seedling immediately attaches itself to the host plant by means of minute suckers.

Yuncker (1932) has made a thorough study of the genus. Engelmann had initiated studies on the American species as early as 1842.

KEY TO THE SPECIES OF Cuscuta IN ILLINOIS

1. Sepals free to base _____ 2
1. Sepals united below into a tube _____ 4
    2. Flowers pedicellate, borne in rather loose cymes or panicles; seeds 1.4–1.5 mm long _____ 1. *C. cuspidata*
    2. Flowers sessile, borne in dense glomerules; seeds 1.7 mm long or longer _____ 3
3. Bracts at base of sepals appressed; lobes of corolla obtuse; seeds 2.5–2.6 mm long_____ 2. *C. compacta*
3. Bracts at base of sepals with recurved tips; lobes of corolla acute; seeds 1.7–1.8 mm long _____ 3. *C. glomerata*
    4. Most of the flowers with 4-lobed corollas _____ 5
    4. Most of the flowers with 5-lobed corollas _____ 7
5. Corolla lobes erect; flowers sessile or on pedicels up to 0.5 mm long __
    _____ 6
5. Corolla lobes inflexed; flowers on pedicels usually at least 1 mm long__
    _____ 6. *C. coryli*
    6. Scales absent, or reduced to minute teeth along the filaments; lobes of corolla acute, about as long as the corolla tube; seeds 1.3–1.4 mm long_____ 4. *C. polygonorum*
    6. Scales toothed from base to apex; lobes of corolla obtuse to subacute, shorter than the corolla tube; seeds 1.6–1.7 mm long _____
    _____ 5. *C. cephalanthi*
7. Lobes of corolla obtuse, erect or spreading_____ 7. *C. gronovii*
7. Lobes of corolla acute, the tips inflexed _____ 8
    8. Lobes of calyx obtuse; pedicels shorter than the flowers _____ 9
    8. Lobes of calyx acute; pedicels as long as or longer than the flowers _____ 10. *C. indecora*
9. Lobes of corolla acuminate; scales about half as long as corolla tube; seeds 1.0–1.2 mm long _____ 8. *C. pentagona*

9. Lobes of corolla acute; scales about as long as corolla tube; seeds 1.5–
1.6 mm long _ 9. *C. campestris*

1. Cuscuta cuspidata Engelm. Bost. Journ. Nat. Hist. 5:224.
1847. *Fig. 51.*

Stems slender, yellowish; flowers in usually rather dense, paniculate cymes; bracts 1–several, ovate, acute, entire, glabrous; sepals
5, free, ovate to suborbicular, cuspidate, entire, to 2.5 mm long;
corolla salverform, white, with 5 oblong to lanceolate, acute,
spreading lobes about half as long as the tube, with 5 narrow, fimbriate scales shorter than the corolla tube; stamens 5, reaching the
base of the sinuses between the corolla lobes; styles slender, longer
than the ovary; stigmas capitate; capsule subglobose, with the withered corolla persistent; seeds 1.4–1.5 mm long.

COMMON NAME: Dodder.
HABITAT: Parasitic primarily on the Asteraceae, particularly *Ambrosia*.
RANGE: Indiana and Wisconsin to Utah, south to Texas
and Louisiana.
ILLINOIS DISTRIBUTION: Scattered but not common in
Illinois.
This is the only species of *Cuscuta* in Illinois with free
sepals and pedicellate flowers. The somewhat similar *C.
compacta* and *C. glomerata* haver sessile flowers. Although there is variation in texture of flowers and number of bracts, these variations are not worthy of names.

The flowers bloom from July to October.

2. Cuscuta compacta Juss. ex Choisy, Mem. Soc. Gen. 9:281,
pl. 4, f. 2. 1841. *Fig. 52.*

Stems rather stout, yellowish-white; flowers sessile in dense clusters; bracts 3–5, orbicular, obtuse, glabrous, serrulate; sepals 5,
free, orbicular, obtuse, glabrous, to 3 mm long; corolla salverform
to urceolate, white, with 5 oblong to ovate, obtuse to subacute,
spreading lobes less than half as long as the tube, with 5 narrow,
fimbriate scales shorter than the corolla tube; stamens 5, reaching
the base of the sinuses between the corolla lobes; styles slightly
shorter than the ovary; stigmas capitate; capsule oblongoid, with
the withered corolla persistent; seeds 2.5–2.6 mm long.

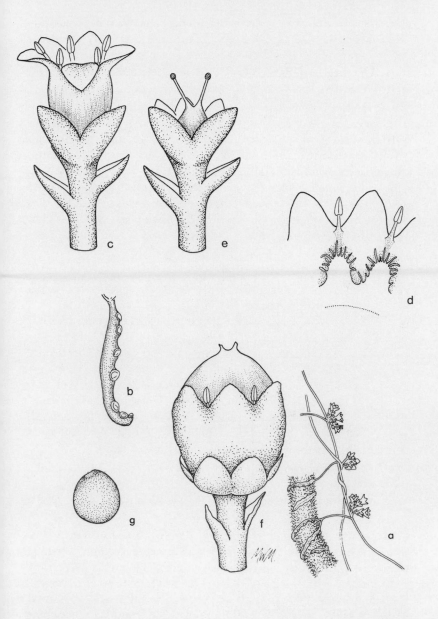

51. *Cuscuta cuspidata* (Dodder). *a.* Habit, on host, × ½. *b.* Stem, with haustoria,
× 5. *c.* Flower, × 10. *d.* Corolla (partly cut away), × 10. *e.* Flower, with corolla
removed, × 10. *f.* Fruit, × 10. *g.* Seed, × 10.

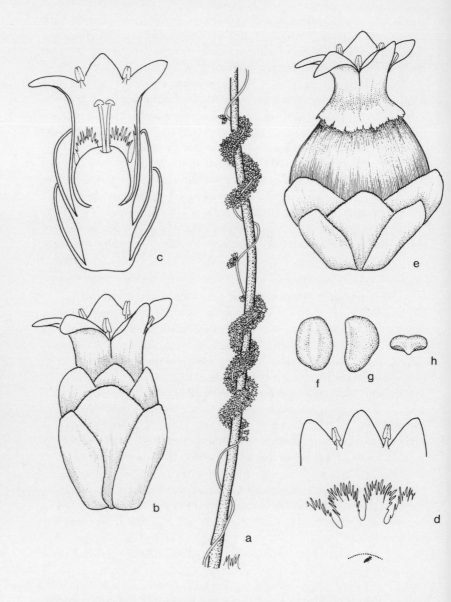

52. *Cuscuta compacta* (Dodder). *a.* Habit, on host, ×½. *b.* Flower, ×10. *c.* Flower (partly cut away), ×10. *d.* Corolla (opened out to show stamens and scales), ×10. *e.* Developing fruit, ×10. *f.* Seed, face-view. ×5. *g.* Seed, side-view, ×5. *h.* Seed, top-view, ×5.

COMMON NAME: Dodder.

HABITAT: Parasitic mostly on shrubby plants in low ground.

RANGE: New Hampshire across Pennsylvania to Oklahoma, south to Texas and Florida.

ILLINOIS DISTRIBUTION: Scattered but not common in the state.

This species is similar to *C. glomerata* in its sessile flowers, but the flowers do not form the dense clusters as the flowers of *C. glomerata* do.

Cuscuta compacta has the largest seeds of any species of *Cuscuta* in the state.

The flowers bloom from July to October.

3. **Cuscuta glomerata** Choisy, Mem. Soc. Gen. 9:280, pl. 4, f. 1. 1841. *Fig. 53.*

Lepidanche compositorum Engelm. Am. Journ. Sci. 43:344. 1842.

Stems slender, yellowish white; flowers sessile in dense clusters, usually completely concealing the stem of the host; bracts 8 or more, lanceolate, acute and recurved at tip, serrulate, glabrous; sepals 5, free, oblong, obtuse, serrulate, glabrous, to 2.5 mm long; corolla oblong-cylindric, white, with 5 oblong to lanceolate, subacute, spreading lobes less than half as long as the tube, with 5 narrow, fimbriate scales shorter than the corolla tube; stamens 5, reaching the base of the sinuses between the corolla lobes; styles at least twice as long as the ovary; stigmas capitate; capsule oblongoid, with the withered corolla persistent; seeds 1.7–1.8 mm long.

COMMON NAME: Dodder.

HABITAT: Primarily in low areas on Asteraceae.

RANGE: Michigan to South Dakota, south to Texas and Mississippi.

ILLINOIS DISTRIBUTION: Occasional and scattered throughout the state.

The sessile flowers grow in compact clusters so densely that the stem of the host is usually obliterated. The flowers are somewhat fragrant. The long styles are also diagnostic for this species. The type was collected by N. Riehl near St. Louis in 1838.

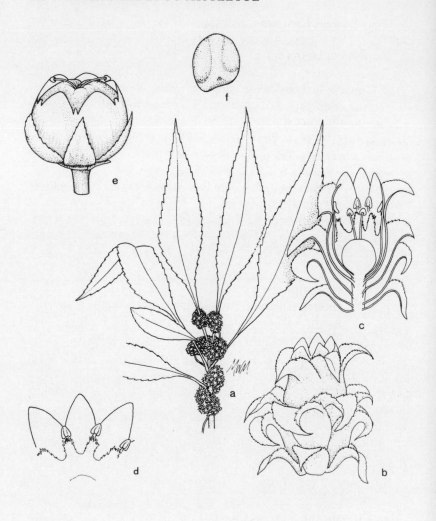

53. *Cuscuta glomerata* (Dodder). *a.* Habit, on leafy host, × ½. *b.* Flower, × 10. *c.* Side-view of flower, × 10. *d.* Corolla (opened out to show stamens and scales), × 10. *e.* Fruit, × 10. *f.* Seed, × 10.

Many species have been recorded as host for *Cuscuta glomerata*.

The flowers bloom from July to October.

4. **Cuscuta polygonorum** Engelm. Am. Journ. Sci. 43:342. 1842. *Fig. 54.*
   Cuscuta chlorocarpa Engelm. in Gray, Man. Bot. 350. 1848.

Stems slender, orange; flowers pedicellate in glomerules; bracts absent; calyx usually 4-lobed, the lobes ovate, obtuse, glabrous, about as long as the corolla tube; corolla campanulate, white, with 4 deltoid, acute, erect lobes about as long as the tube, with minute toothlike scales, or the scales absent; stamens 5, arising from the sinuses between the corolla lobes; styles shorter than the ovary; stigmas capitate; capsule globose, with the withered corolla persistent; seeds 1.3–1.4 mm long.

COMMON NAME: Dodder.
HABITAT: Parasitic on various herbs in moist ground.
RANGE: Quebec to Minnesota, south to Oklahoma, Tennessee, and Maryland.
ILLINOIS DISTRIBUTION: Occasional and scattered throughout the state.

This species occurs in low areas, particularly on species of *Polygonum,* as the specific epithet indicates.

This is the only species of *Cuscuta* in Illinois with obsolete scales.

The type was collected at St. Louis by F. Lindheimer in August, 1839,

The flowers open from July to September.

5. **Cuscuta cephalanthi** Engelm. Am. Journ. Sci. 43:336. 1842. *Fig. 55.*

Stems coarse, yellow; flowers pedicellate in glomerules; bracts absent; calyx usually 4-lobed, the lobes ovate, obtuse, glabrous, much shorter than the corolla tube; corolla cylindric-campanulate, white, with 4 ovate, obtuse, spreading lobes, about half as long as the tube, with 5 narrowly oblong, toothed scales; stamens 5, included; styles about as long as the ovary; stigmas capitate; capsule globose, with the withered corolla persistent; seeds 1.6–1.7 mm long.

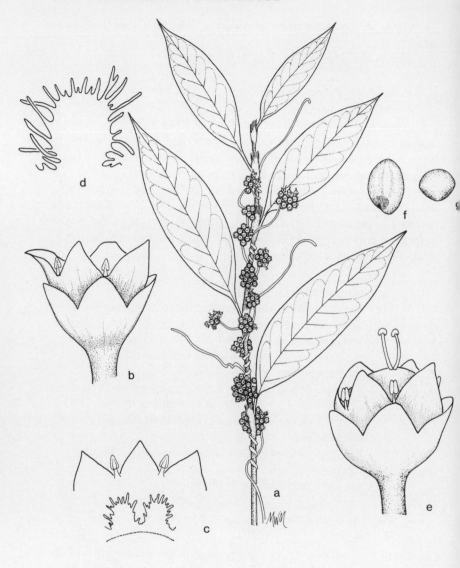

54. *Cuscuta polygonorum* (Dodder). *a.* Habit, on leafy host, ×½. *b.* Flower, ×10. *c.* Corolla (opened out to show stamens and scales), ×10. *d.* Scale, ×25. *e.* Developing fruit, ×10. *f.* Seed, face-view, ×10. *g.* Seed, top-view, ×10.

55. *Cuscuta cephalanthi* (Dodder). *a*. Habit, on leafy host, ×½. *b*. Flower, ×10. *c*. Corolla (opened out to show stamens and scales), ×10. *d*. Fruit, ×10. *e*. Seed, ×10.

COMMON NAME: Dodder.

HABITAT: Parasitic on various shrubs and herbs in moist ground.

RANGE: Nova Scotia to British Columbia, south to New Mexico and Virginia.

ILLINOIS DISTRIBUTION: Occasional and scattered throughout the state.

This species shows a preference for shrubby species that grow in low ground.

The diagnostic features of *C. cephalanthi* are the 4-lobed, erect corollas with toothed scales.

The type was collected by Engelmann from St. Louis in 1841.

The flowers bloom from August to October.

6. Cuscuta coryli Engelm. Am. Journ. Sci. 43:337. 1842. *Fig. 56.*

Cuscuta inflexa Engelm. Trans. St. Louis Acad. 1:502. 1859.

Stems coarse, yellow; flowers pedicellate in cymes; bracts absent; calyx usually 4-lobed, the lobes lanceolate, acute, glabrous, about as long as the corolla tube; corolla campanulate, white, with 4 lanceolate, acute but incurved-tipped, erect bracts about as long as the tube, with 5 small, narrow, bifid scales; stamens 5, included; styles shorter than the ovary; stigmas capitate; capsule oblongoid, with the withered corolla persistent; seeds 1.4–1.5 mm long.

COMMON NAME: Dodder.

HABITAT: Parasitic on various shrubs and tall herbs.

RANGE: Connecticut to Montana, south to Arizona and North Carolina.

ILLINOIS DISTRIBUTION: Occasional and scattered throughout the state.

Cuscuta coryli is unique in having four corolla lobes with inflexed tips. Also unique are the narrow, bifid scales.

The type was collected by George Engelmann from St. Louis in 1841.

The flowers appear from July to October.

7. Cuscuta gronovii Willd. ex Roem. & Schultes, Syst. 6:205. 1820.

Stems slender, yellow-orange; flowers pedicellate in paniculate cymes; bracts absent; calyx 5-lobed, the lobes ovate, obtuse, gla-

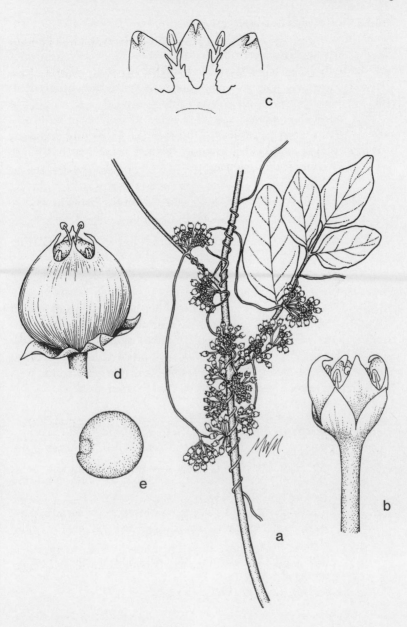

56. *Cuscuta coryli* (Dodder). *a.* Habit, on leafy host, ×½. *b.* Flower, ×10. *c.* Corolla (opened out to show stamens and scales), ×10. *d.* Fruit, ×10. *e.* Seed, ×10.

brous, shorter than to about as long as the corolla tube; corolla campanulate to cylindrical, white, with 5 ovate, obtuse, spreading lobes about as long as the tube, with 5 narrow scales fimbriate at the apex and toothed at the base; stamens 5, included; styles shorter than the ovary; stigmas capitate; capsule globose, with the withered corolla persistent; seeds 1.4–1.5 mm long.

Two rather distinct varieties of *C. gronovii* can be found in Illinois. In their extreme conditions, the varieties look quite different. Engelmann had recognized them as distinct species at one time, but later reduced them to varieties.

KEY TO THE VARIETIES OF C. gronovii IN ILLINOIS

1. Calyx lobes shorter than the corolla; corolla cylindrical_____
   _____ 7a. *C. gronovii* var. *gronovii*
1. Calyx lobes about as long as the corolla; corolla campanulate _____
   _____ 7b. *C. gronovii* var. *latiflora*

7a. Cuscuta gronovii Willd. var. **gronovii** *Fig. 57a–e.*
Cuscuta vulvivaga Engelm. Am. Journ. Sci. 43:338. 1842.
Cuscuta gronovii Willd. var. *vulvivaga* (Engelm.) Engelm. Trans. St. Louis Acad. 1:508. 1859.
Calyx lobes shorter than the corolla; corolla cylindrical.

COMMON NAME: Dodder.
HABITAT: Parasitic on herbs growing in moist areas.
RANGE: Quebec to Manitoba, south to Arizona and Florida.
ILLINOIS DISTRIBUTION: Common throughout the state.
This is the more common variety of *C. gronovii* in Illinois. Apparently unaware of Willldenow's *C. gronovii*, Engelmann described the same species in 1842 as *C. vulvivaga*. In 1859, he later called the plant *C. gronovii* var. *vulvivaga*. I believe that Engelmann's *C. vulvivaga* is synonymous with Willldenow's *C. gronovii*.
The flowers bloom from July to October.

7b. Cuscuta gronovii Willd. var. **latiflora** Engelm. Trans. St. Louis Acad. 1:508. 1859. *Fig. 57f.*
Cuscuta saururi Engelm. Am. Journ. Sci. 43:339. 1842.

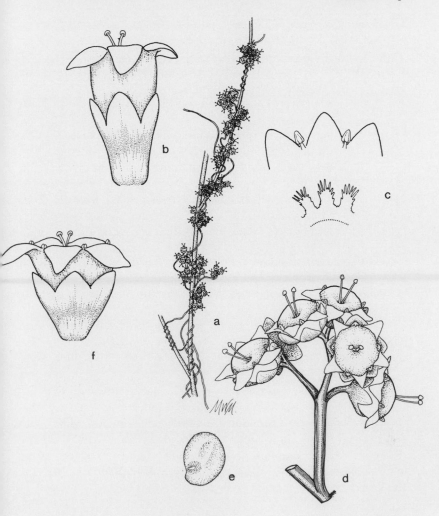

57. *Cuscuta gronovii* (Dodder). *a.* Habit, on host, ×½. *b.* Flower, ×10. *c.* Corolla (opened out to show stamens and scales), ×10. *d.* Cluster of fruits, ×3½. *e.* Seed, ×10. var. *latiflora* (Dodder). *f.* Flower, ×10.

Cuscuta gronovii Willd. var. *saururi* (Engelm.) MacM. Geol. & Nat. Hist. Surv. Minn. 1:430. 1892.

Calyx lobes about as long as the corolla; corolla campanulate.

COMMON NAME: Dodder.

HABITAT: Parasitic on herbs growing in low, moist areas.

RANGE: New Jersey to Missouri, south to Texas.

ILLINOIS DISTRIBUTION: Not common in Illinois.

This variety is much less common in Illinois than var. *gronovii*.

A specimen collected by Geyer in 1841 from St. Clair County was chosen by Engelmann as his type for *C. saururi*. It is not different from var. *latiflora*.

8. **Cuscuta pentagona** Engelm. Am. Journ. Sci. 43:340. 1842. *Fig. 58.*

Cuscuta arvensis Beyr. ex Hook. Fl. Bor. Am. 2:77. 1838, in synon.

Cuscuta pentagona Engelm. var. *microcalyx* Engelm. Am. Journ. Sci. 45:76. 1845.

Cuscuta pentagona Engelm. var. *typica* Yuncker, Ill. Biol. Mon. 6:50. 1921.

Stems very slender, pale yellow; flowers short-pedicellate in glomerules; bracts absent; calyx 5-lobed, the lobes ovate, obtuse, glabrous, as long as the corolla tube; corolla campanulate, white, with 5 lanceolate, acuminate-incurved, spreading lobes, with 5 narrow, fimbriate scales; stamens 5, included or slightly exserted; styles shorter than the ovary; stigmas capitate; capsule globose, with the withered corolla persistent; seeds 1.0–1.2 mm long.

COMMON NAME: Dodder.

HABITAT: Parasitic on various herbs, particularly legumes.

RANGE: Massachusetts to Montana and California, south to Texas and Florida; Mexico.

ILLINOIS DISTRIBUTION: Occasional throughout the state.

Variety *microcalyx* was described by Engelmann from an Illinois collection. It does not, however, seem to be worthy of recognition.

The flowers of *C. pentagona* open from June to October.

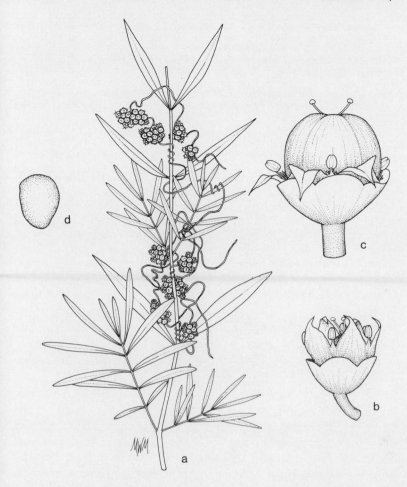

58. *Cuscuta pentagona* (Dodder). *a.* Habit, on leafy host, × ½. *b.* Flower, × 10. *c.* Fruit, × 10. *d.* Seed, × 10.

9. **Cuscuta campestris** Yuncker, Mem. Torrey Club 18:138. 1932. *Fig. 59.*

Cuscuta pentagona Engelm. var. *calycina* Engelm. Am. Journ. Sci. 45:76. 1845.

Stems slender, pale yellow; flowers short-pedicellate, in glomerules; bracts absent; calyx 5-lobed, the lobes ovate, obtuse, nearly as long as the corolla tube; corolla campanulate, white, with 5 lance-ovate, acute-incurved, spreading lobes, with 5 oblong, fimbriate

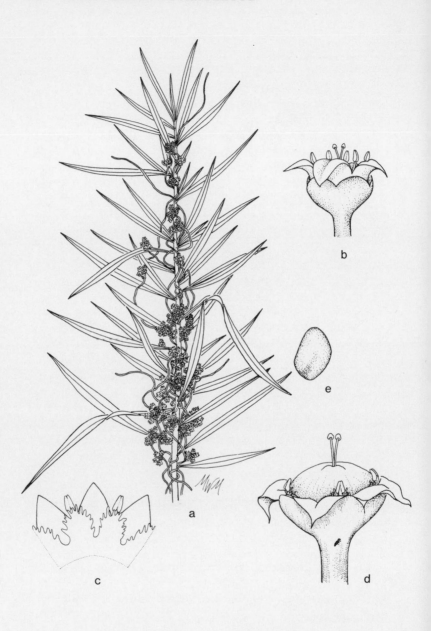

59. *Cuscuta campestris* (Dodder). *a.* Habit, on leafy host, ×½. *b.* Flower, ×10. *c.* Corolla (opened out to show stamens and scales), ×10. *d.* Fruit, ×10. *e.* Seed, ×10.

scales nearly as long as the corolla tube; stamens 5, included; styles about as long as the ovary; stigmas capitate; capsule globose, with the withered corolla persistent; seeds 1.5–1.6 mm long.

 COMMON NAME: Dodder.
HABITAT: Parasitic on various herbs.
RANGE: Massachusetts to Washington, south to California, Texas, and South Carolina; Canada; Mexico; tropical America.
ILLINOIS DISTRIBUTION: Scattered but uncommon in Illinois.
This species is very similar to *C. pentagona*, differing by its longer scales and larger seeds. The calyx lobes generally do not overlap as they do in *C. pentagona*.

Engelmann (1845) originally described this plant as *C. pentagona* var. *calycina*, and this was followed by Yuncker in 1921. Later, Yuncker (1932) raised the rank of this taxon to species.

The flowers bloom from June to October.

10. **Cuscuta indecora** Choisy, Mem. Soc. Gen. 9:278. 1841.

Stems rather coarse, orange-yellow; flowers pedicellate, in cymose panicles; bracts absent; calyx 5-lobed, the lobes ovate to lanceolate, acute, glabrous, half as long as to equaling the corolla tube; corolla campanulate, white, papillate, with 5 deltoid, acute-incurved, spreading lobes,with 5 oblong, short-fimbriate scales; stamens 5, included or barely exserted; styles about as or slightly longer than the ovary; stigmas capitate; capsule oblongoid, with the withered corolla persistent; seeds 1.6–1.7 mm long.

Two varieties may be distinguished in Illinois.

KEY TO THE VARIETIES OF C. indecora IN ILLINOIS

1. Calyx lobes about half as long as the corolla tube _____ _____ 10a. *C. indecora* var. *indecora*
1. Calyx lobes about as long as the corolla tube _____ _____ 10b. *C. indecora* var. *neuropetala*

10a. **Cuscuta indecora** Choisy var. **indecora** *Fig. 60a–e.*
Calyx lobes about half as long as the corolla tube.

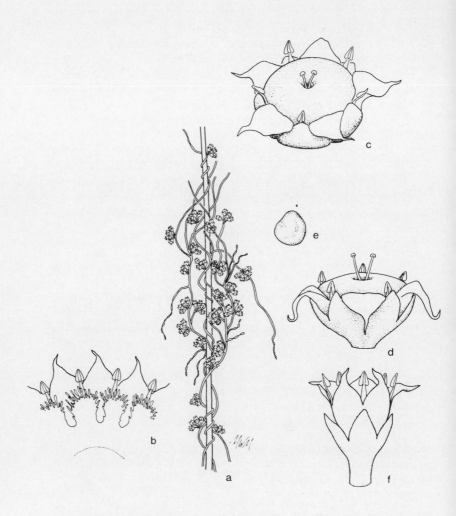

60. *Cuscuta indecora* (Dodder). *a.* Habit, × ½. *b.* Corolla (opened out to show stamens and scales), × 10. *c, d.* Developing fruits, × 10. *e.* Seed, × 10. var. *neuropetala*
(Dodder). *f.* Flower, × 10.

COMMON NAME: Dodder.

HABITAT: Parasitic on various herbs.

RANGE: Virginia to Minnesota and Colorado, south to Texas and Florida; West Indies.

ILLINOIS DISTRIBUTION: Apparently confined to the southern two-thirds of the state.

This variety and the following are distinctive by their papillate corolla. In var. *indecora,* the calyx lobes are conspicuously short.

The flowers bloom from July to September.

10b. Cuscuta indecora Choisy var. **neuropetala** (Engelm.) Hitchc. Contr. U. S. Nat. Herb. 3:549. 1896. *Fig. 6of.*
Cuscuta neuropetala Engelm. Am. Journ. Sci. 45:76. 1845.

Calyx lobes about as long as the corolla tube.

COMMON NAME: Dodder.

HABITAT: Parasitic on various herbs growing in low areas.

RANGE: Illinois and Minnesota to California, south to Texas and Florida; Mexico.

ILLINOIS DISTRIBUTION: Rare in the southern half of the state.

POLEMONIACEAE–PHLOX FAMILY

Annual or perennial herbs, rarely shrubs or small trees; leaves alternate or opposite, simple or compound, without stipules; inflorescence mostly cymose; flowers perfect, actinomorphic; calyx 5-lobed; corolla 5-parted; stamens 5, attached to the corolla tube; disk usually present; pistil 1, the ovary superior, 3-locular, with 1–several ovules on each axile placenta; stigmas 3; fruit a capsule.

This family is comprised of about thirteen genera and 275 species native to North and South America.

The chief diagnostic characters of the family are the trilocular ovary with numerous ovules and the united sepals and united petals.

Several garden ornamentals are found in the family, notably in the genera *Polemonium* (valerian), *Ipomopsis* (standing cypress), *Phlox* (sweet william and phlox), and *Cobaea.*

KEY TO THE GENERA OF Polemoniaceae IN ILLINOIS

1. Leaves pinnately compound or deeply pinnatifid _____ 2
1. Leaves simple, entire _____ 4
   2. Leaves pinnately compound with lanceolate to oval leaflets; flowers blue, campanulate, to 1.5 cm long_____ 1. *Polemonium*
   2. Leaves deeply pinnatifid, the segments filiform; flowers red or pink, narrowly funnelform, at least 2.5 cm long _____ 3
3. Flowers salverform, each one subtended by a bract; seeds long, slender, curved _____ 2. *Ipomopsis*
3. Flowers funnelform, only the clusters subtended by a bract; seeds spherical _____ 3. *Gilia*
   4. All leaves alternate _____ 4. *Collomia*
   4. At least the lower cauline leaves opposite _____ 5
5. Usually all leaves opposite; calyx actinomorphic; corolla 1.5 cm long or longer _____ 5. *Phlox*
5. Uppermost leaves alternate; calyx slightly zygomorphic; corolla 8–12 mm long_____ 6. *Microsteris*

1. *Polemonium* L. –Greek Valerian

Annual or perennial herbs; leaves alternate, pinnately compound; flowers perfect, actinomorphic, in cymose panicles or thyrses; calyx campanulate, 5-lobed; corolla campanulate to funnelform, 5-lobed; stamens 5, attached to the corolla; ovary superior, 3-locular; fruit a 3-valved capsule.

Polemonium is a genus of nearly fifty species, mostly native to western North America.

Only the following species occurs in Illinois.

1. **Polemonium reptans** L. Syst. ed. 10, 2:925. 1759. *Fig. 61.*
Polemonium reptans L. var. *villosum* E. L. Braun, Rhodora 42:50. 1940.
Polemonium reptans L. f. *villosum* (E. L. Braun) Wherry, Am. Midl. Nat. 27:753. 1942.
Perennial herbs; stems widely spreading, much branched, glabrous or rarely villous, to 45 cm long; lowermost leaves with 11–17 leaflets, the leaflets lanceolate to oblong to ovate, acute at the apex, subcuneate at the sessile or nearly sessile base, glabrous or rarely villous, to 3 cm long, the uppermost leaves with 3–5 leaflets or

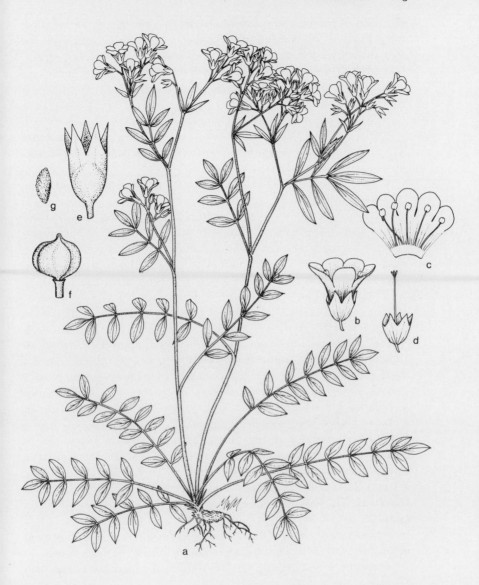

61. *Polemonium reptans* (Jacob's Ladder). *a.* Habit, × ½. *b.* Flower, × 1½. *c.* Corolla (opened out to show stamens), × 1½. *d.* Flower (corolla and stamens removed), × 1½. *e.* Capsule (enclosed by calyx), × 2. *f.* Capsule (calyx removed), × 3. *g.* Seed, × 5.

even simple; flowers in corymbs, to nearly 2 cm across, on pedicels to 1.5 cm long; calyx campanulate, green, glabrous or nearly so, the 5 lobes deltoid to lanceolate, more or less acute; corolla campanulate, blue, 5-lobed; stamens 5, included; capsule ovoid, glabrous, 3-valved, with about 3 seeds.

COMMON NAME: Jacob's Ladder.
HABITAT: Moist or sometimes dry woods; prairies; fens.
RANGE: New York to Minnesota, south to Oklahoma and Georgia.
ILLINOIS DISTRIBUTION: Common throughout the state.
This is one of our more attractive spring wild flowers. Its common habitat is woodlands, but occasional specimens may be found in calcareous fens and prairies.
Specimens with villous stems and leaves are rarely encountered.
The flowers bloom from April to June.

2. *Ipomopsis* Michx.–Standing Cypress

Mostly perennials or biennials, less commonly annuals; leaves alternate, pinnatisect, often some of them in a basal rosette; flowers mostly in cymes, rarely racemes, each subtended by a bract; calyx united, the lobes mucronate; corolla salverform, 5-lobed; stamens 5, attached to the corolla; pistil 1, the ovary superior, 3-locular; fruit a capsule; seeds long, slender, curved.

In his recognition of the genus *Ipomopsis*, Grant (1956) attributed twenty-three species to the genus. They are found in the Rocky Mountains and adjacent plains, south to Texas and Mexico, west to the Pacific Coast and Southwest, east to Florida and the Carolinas. There is one species in the Argentine Andes and Patagonia.

Ipomopsis not only differs from *Gilia* in several morphological characters, but also in having seven pairs of large chromosomes.

Only the following escape from cultivation has been found in Illinois.

1. **Ipomopsis rubra** (L.) Wherry, Bartonia 18:56. 1936. *Fig. 62.*
Polemonium rubrum L. Sp. Pl. 163. 1753.
Gilia rubra (L.) Heller, Contr. Herb. Frankl. & Marsh. Coll. 1:81. 1895.

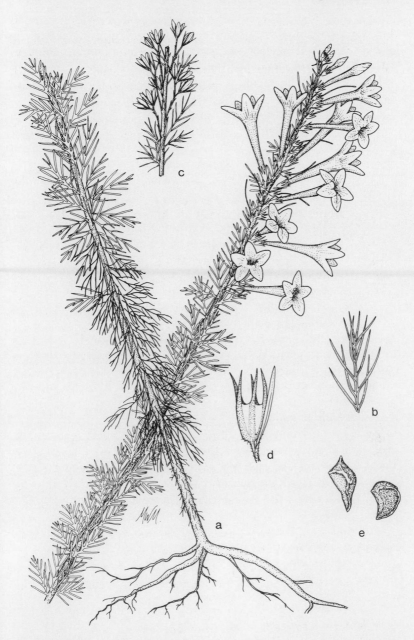

62. *Ipomopsis rubra* (Standing Cypress). *a*. Habit, × ½. *b*. Leaf, × 1. *c*. Upper part of plant, × ½. *d*. Fruit, × 2½. *e*. Seeds, × 5.

Perennial or biennial herbs; stems erect, simple, glabrous, to 1 m tall; leaves alternate, pinnately divided into numerous filiform to linear segments, the segments to 2.5 cm long, glabrous; flowers several in a narrow, terminal thyrse or panicle, nearly sessile, each subtended by a bract longer than the calyx; calyx campanulate, green, 5-lobed, the lobes deltoid-triangular; corolla salverform, usually scarlet, to 3.5 cm long, 5-lobed, the lobes ovate, spreading; stamens 5, mostly included; capsule ovoid, with numerous seeds.

COMMON NAME: Standing Cypress.

HABITAT: Roadsides.

RANGE: Native of the southern United States; adventive as a garden escape in the eastern half of the country.

ILLINOIS DISTRIBUTION: Known from Clark, Putnam, Washington, and Woodford counties; reported by Wherry (1936) from Winnebago County, but not verified in this study.

This is a frequently grown garden ornamental that occasionally escapes from gardens but seldom persists. It frequently is known as *Gilia rubra*.

The numerous filiform leaf segments and the narrow, salverform corolla are distinctive.

The flowers bloom from June to August.

3. *Gilia* R. & P.–Gilia

Mostly annual, less commonly biennial or perennial herbs; leaves alternate or opposite, simple or compound; flowers solitary or in cymes or panicles, only the flower clusters bracteate; calyx campanulate, 5-lobed; corolla funnelform, 5-lobed; stamens 5, attached to the corolla; pistil 1, the ovary superior, 3-locular; fruit a capsule; seeds spherical.

With the removal of several species to the closely related genus *Ipomopsis*, *Gilia* contains about seventy-five species, native primarily to western North America.

One species, an escape from cultivation, has been found in Illinois.

1. **Gilia capitata** Sims, Bot. Mag. 2698. 1826. *Fig. 63.*

Annual; stems erect, usually glabrous, to 80 cm tall; leaves much divided, the ultimate segments up to 1 mm broad; flowers borne in

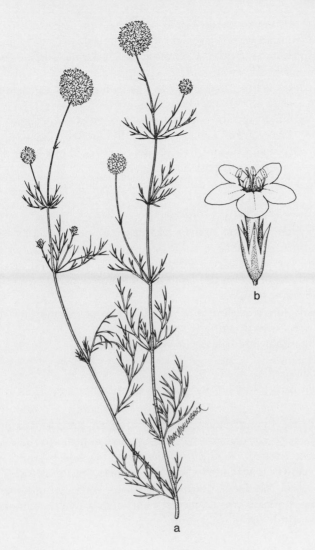

63. Gilia capitata (Gilia). *a.* Habit, × ½. *b.* Flower, × 3.

terminal heads up to 4 cm thick, with each head bearing 50 or more flowers; calyx united, 5-lobed, up to 4 mm long; corolla 5-lobed, pale violet, up to 8 mm long; capsule subglobose, up to 4 mm in diameter; seeds angular, 1.5–2.0 mm long.

COMMON NAME: Gilia.

HABITAT: Cultivated soil (in Illinois).

RANGE: British Columbia to California; Idaho; adventive elsewhere.

ILLINOIS DISTRIBUTION: Known only as a garden escape in Putnam County.

Gilia capitata is an infrequently grown and highly variable garden ornamental in Illinois. Its collection from Illinois adjacent to a garden casts doubt as to its spontaneity in Illinois.

The flowers bloom from June to September.

4. *Collomia* Nutt.–Collomia

Annual or biennial herbs; leaves alternate, simple; flowers perfect, actinomorphic, in dense cymes or capitate clusters; calyx cup-shaped, 5-lobed; corolla funnelform or salverform, 5-lobed; stamens 5, unequally inserted on the tube of the corolla; pistil 1, the ovary superior; fruit a capsule.

Collomia is comprised of about fifteen species, most of them native to the western United States.

Only the following species occurs in Illinois.

1. Collomia linearis Nutt. Gen. 1:126. 1818. *Fig. 64.*

Gilia linearis (Nutt.) Gray, Proc. Am. Acad. 17:233. 1882.

Annual herb from fibrous roots; stems spreading to ascending, simple or branched, viscid-pubescent, to 45 cm long; leaves linear to lanceolate, acute to acuminate at the apex, cuneate to the nearly sessile base, entire, viscid-pubescent, to 5 cm long; flowers in a terminal cluster subtended by leafy bracts; calyx cup-shaped, green, 5-lobed, the lobes deltoid-lanceolate, acute; corolla funnelform, pale purple to white, 1.0–1.5 cm long, 5-lobed; stamens 5, included; capsule ovoid, to 5 mm long.

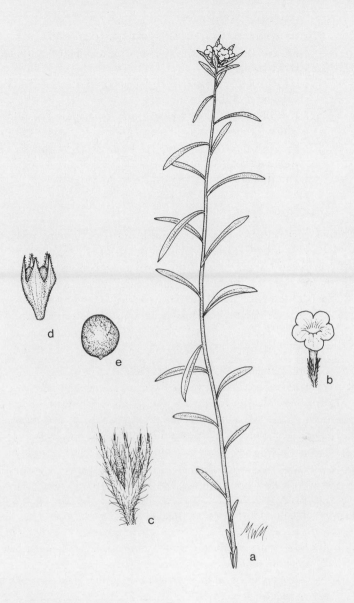

64. Collomia linearis (Collomia). *a.* Habit, ×½. *b.* Flower, ×1½. *c.* Calyx, ×5. *d.* Fruit, ×2½. *e.* Seed, ×2½.

COMMON NAME: Collomia.

HABITAT: Along railroads; dry waste ground.

RANGE: Native to the western United States; adventive eastward.

ILLINOIS DISTRIBUTION: Occasional in the northern one-third of the state.

The flowers of this species vary from pale lavender to white.

It is the only species of Polemoniaceae in Illinois with all leaves simple and alternate.

The flowers bloom from June to August.

5. *Phlox* L. –Phlox

Perennial herbs; leaves opposite, simple, entire; flowers mostly in cymes; calyx 5-lobed; corolla salverform, 5-lobed; stamens 5, unequally inserted on the tube of the corolla, usually included; pistil 1, the ovary superior, 3-locular, with 1–4 ovules per locule.

Phlox is a genus of about seventy species, all but one native to North America. Several species are grown as garden ornamentals.

KEY TO THE SPECIES OF Phlox IN ILLINOIS

1. Petals deeply notched or emarginate at the tip _____ 2
1. Petals not notched or emarginate at the tip _____ 4
    2. Leaves lanceolate to narrowly ovate; plants erect or ascending, not diffusely branched; petals emarginate _____ 3. *P. divaricata*
    2. Leaves linear to subulate; plants diffuse and much branched; petals deeply notched (sometimes emarginate in a rare subspecies of *P. bifida*) _____ 3
3. Leaves linear, usually without fascicles of smaller leaves in the axils ___ _____ 1. *P. bifida*
3. Leaves subulate, with fascicles of smaller leaves in the axils _____ _____ 2. *P. subulata*
    4. Calyx lobes longer than calyx tube; stamens and style about half as long as corolla tube _____ 5
    4. Calyx lobes equaling or shorter than calyx tube; stamens and style about as long as corolla tube _____ 6
5. Leaves lanceolate to elliptic to narrowly ovate; plants often with sterile leafy shoots _____ 3. *P. divaricata*
5. Leaves linear to narrowly lanceolate; plants without sterile leafy shoots_____ 4. *P. pilosa*

6. Leaves with conspicuous lateral veins and reticulations; calyx teeth subulate _____ 5. *P. paniculata*
6. Leaves without conspicuous lateral veins and reticulations; calyx teeth lanceolate _____ 7
7. Flowers in panicles usually longer than broad; stems usually purple-spotted_____ 6. *P. maculata*
7. Flowers in corymbs usually nearly as broad as long; stems green or purplish, not spotted _____ 8
8. Calyx to 7.5 mm long _____ 7. *P. glaberrima*
8. Calyx 9–12 mm long _____ 8. *P. carolina*

1. Phlox bifida Beck, Am. Journ. Sci. 11:170. 1826.

Perennial herb from a somewhat woody base; stems diffuse, mat forming, much branched, wiry, to 20 cm long, glabrous to densely pubescent, the pubescence sometimes glandular; leaves linear to linear-lanceolate, acute at the apex, cuneate to the sessile base, entire, glabrous or pubescent, to 4 cm long, without fascicles of other leaves in the axils; flowers solitary or in cymes, axillary, on glabrous or glandular-pubescent pedicels to 3 cm long; calyx 5-lobed, the lobes subulate, longer than the calyx tube, pubescent; corolla salverform, pale purple to nearly white, the 5 lobes sometimes emarginate but usually notched to about the middle, with conspicuous dark eye-markings; style to 1 cm long, longer than the calyx lobes; capsule oblongoid, to 3 mm long.

Two subspecies may be distinguished in Illinois.

KEY TO THE SUBSPECIES OF P. bifida IN ILLINOIS

1. Some of the pubescence glandular; petals notched to about the middle _____ 1a. *P. bifida* ssp. *bifida*
1. None of the pubescence glandular; petals emarginate or notched only about one-fourth their length _____ 1b. *P. bifida* ssp. *stellaria*

1a. Phlox bifida Beck ssp. bifida *Fig. 65a–f.*

Some of the pubescence glandular; petals notched to about the middle.

65. *Phlox bifida* (Cleft Phlox). *a.* Habit, ×½. *b.* Leaf, ×1. *c.* Leaf margin showing glandular hairs, ×10. *d.* Flower, ×1½. *e.* Fruit, ×2. *f.* Seed, ×7½. ssp. *stellaria* (Cleft Phlox). *g.* Leaf variation, ×½. *h.* Flower, ×1½.

COMMON NAME: Cleft Phlox.
HABITAT: Dry, rocky woods; cliffs.
RANGE: Michigan to Iowa, south to eastern Oklahoma and Kentucky.
ILLINOIS DISTRIBUTION: Throughout the state, except for some of the central counties.
This is the common subspecies of cleft phlox in the state. Variation is exhibited in flower color, depth of the notch of the corolla lobes, and degree of pubescence.
In some areas, this subspecies may grow in dense mats. It flowers from late March to June.

1b. **Phlox bifida** Beck ssp. **stellaria** (Gray) Wherry, Castanea 16:99. 1951. *Fig. 65g, h.*
Phlox stellaria Gray, Proc. Am. Acad. 8:252. 1870.
Phlox stellaria Gray var. *cedaria* Brand, Pflanzenr, 4, Fam. 250:75. 1907.
Phlox bifida Beck var. *stellaria* (Gray) Wherry, Bartonia 11:34. 1929.
Phlox bifida Beck var. *cedaria* (Brand) Fern. Rhodora 51:78. 1949.
None of the pubescence glandular; petals emarginate or notched only about one-fourth their length.

COMMON NAME: Cleft Phlox.
HABITAT: Sandy soil; limestone cliffs.
RANGE: Southern Indiana to southern Missouri, south to Arkansas and Tennessee. There is also a single record from Michigan.
ILLINOIS DISTRIBUTION: Known only from Monroe County.
This uncommon subspecies is distinguished by the absence of glandular hairs and its shallowly notched or emarginate petals.
This distinction of this subspecies has been questioned by a number of botanists. It appears that, at least on some specimens of typical ssp. *bifida*, the glands of the hairs disappear or are knocked off as the plants mature. The type collection from Kentucky has no pubescence whatsoever. Cypher, who is studying the

Phlox bifida complex as this manuscript goes to press, has found eglandular specimens only from Monroe County.

Subspecies *stellaria* flowers from early March to early May.

2. Phlox subulata L. Sp. Pl. 152. 1753. *Fig. 66.*

Perennial herb from a somewhat woody base; stems much branched, mat forming, glabrous or pubescent, the pubescence rarely glandular; leaves linear-subulate, to 2 cm long, entire, glabrous or pubescent, with fascicles of smaller leaves in the axils; flower solitary or in cymes, axillary, on usually glandless pedicels; calyx 5-lobed, the lobes subulate, longer than the calyx tube, glandless but usually pubescent; corolla salverform, purple (in Illinois), rose, or white, the 5 lobes emarginate for 1–2 cm; stamens exserted; style to 1.2 cm long, longer than the calyx lobes; capsule ovoid, to 4 mm long.

COMMON NAME: Moss Pink.

HABITAT: In sandy soil along a road (in Illinois).

RANGE: New York to Michigan, south to Tennessee and North Carolina; sometimes spread from cultivation elsewhere.

ILLINOIS DISTRIBUTION: Mason County: sand timber understory about 17 km southwest of Bath, April 25, 1981, *Mike Mibb 203b.*

Moss pink is a handsome, early-blooming species that is commonly grown as an ornamental, particularly in the southern half of the state. The collection from Mason County cited above is some distance from the nearest cultivated colony.

The flowers bloom from mid-March to early May.

3. Phlox divaricata L. ssp. laphamii (Wood) Wherry, Baileya 4:97. 1956. *Fig. 67.*

Phlox laphamii Wood, Obj. Less. Bot. 265. 1863.

Perennial herb; stems ascending or spreading, often rooting at the lowest nodes, simple or branched, viscid-pubescent, to 45 cm tall; leaves of sterile shoots elliptic, obtuse at the apex, subcuneate at the base, to 4 cm long; leaves of the fertile shoots lanceolate to lance-ovate, acute at the apex, rounded to subcordate at the sessile base, to 5 cm long; flowers borne in cymes, on slender, divergent, glandular-pubescent pedicels; calyx deeply 5-lobed, the lobes sub-

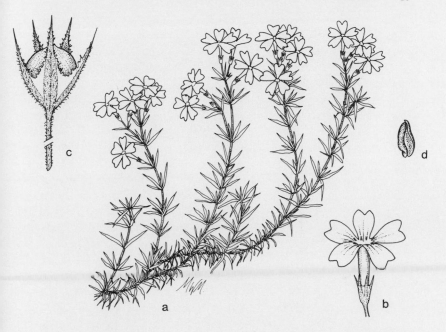

66. *Phlox subulata* (Moss Pink). *a.* Habit, × ½. *b.* Flower, × 1½. *c.* Fruit, × 5. *d.* Seed, × 5.

ulate, much longer than the calyx tube, pubescent, green; corolla salverform, bluish purple to violet to white (rarely green), the 5 lobes entire or emarginate, the notch never more than 1 mm deep, the lobes about as long as the corolla tube; stamens 5, included; style to 3 mm long, included; capsule oblongoid, glabrous, to 4 mm long.

COMMON NAME: Blue Phlox.

HABITAT: Woods.

RANGE: Wisconsin to South Dakota, south to eastern Texas and western Georgia.

ILLINOIS DISTRIBUTION: Common throughout the state.

This is a very common wild flower throughout the entire state. There is some variation in flower color.

Typical ssp. *divaricata* has the notch of each corolla lobe more than 1 mm deep. The transition zone between ssp. *divaricata* and ssp. *laphamii* apparently lies

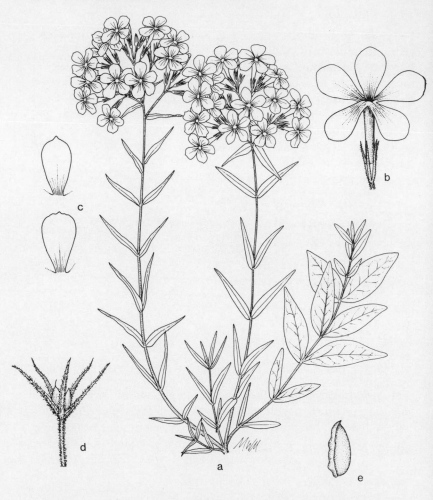

67. *Phlox divaricata* ssp. *laphamii* (Blue Phlox). *a.* Habit, with sterile shoot, × ½. *b.* Flower, × 1½. *c.* Corolla lobe variations, × 1½. *d.* Fruit, × 2. *e.* Seed, × 7½.

just a short distance east of Illinois. Some specimens of *P. divaricata* from the extreme eastern part of Illinois exhibit emarginate corolla lobes, and at least two specimens from Kankakee County have notches about 1 mm deep.

Specimens of ssp. *laphamii* with diminutive green corollas have been seen from Jackson and Randolph counties.

The attractive flowers bloom from April to June. The sterile shoots often persist through most of the winter.

4. Phlox pilosa L. Sp. Pl. 152. 1753.

Perennial herb; stems ascending to erect, simple or branched, glandular-pubescent, pubescent, or glabrous, to 50 cm tall; leaves linear to lanceolate, acute to acuminate at the apex, cuneate or rounded at the base, pubescent to glabrous, to 10 cm long; flowers borne in dense cymes, on short, often glandular, pedicels; calyx deeply 5-lobed, the lobes subulate, longer than the calyx tube, pubescent to glabrous, green; corolla salverform, rose or pink, rarely white, 5-lobed; stamens 5, included; style to 3 mm long, included; capsule ovoid, to 4 mm long.

Levin (1966) has made a thorough study of the *P. pilosa* complex.

KEY TO THE SUBSPECIES OF **P. pilosa** IN ILLINOIS

1. Stems, leaves, and calyx pubescent _____ 2
1. Stems, leaves, and calyx glabrous or nearly so _____
   _____ 4c. *P. pilosa* ssp. *sangamonensis*
   2. Pubescence glandular _____ 4a. *P. pilosa* ssp. *pilosa*
   2. Pubescence eglandular _____ 4b. *P. pilosa* ssp. *fulgida*

4a. Phlox pilosa L. ssp. **pilosa** *Fig. 68a–3.*
Phlox aristata Michx. Fl. Bor. Am. 1:144. 1803.
Phlox aristata Michx. var. *virens* Michx. Fl. Bor. Am. 1:144. 1803.
Phlox argillacea Clute & Ferriss, Am. Bot. 17:74. 1911.
Phlox pilosa L. var. *virens* (Michx.) Wherry, Bartonia 12:47. 1931.
Stems, leaves, and calyx glandular-pubescent.

COMMON NAME: Downy Phlox.
HABITAT: Dry, rocky woods; prairies.
RANGE: Connecticut to Michigan to Kansas, south to Texas and Florida; Ontario.
ILLINOIS DISTRIBUTION: Occasional throughout the state.
This subspecies is more common in Illinois than either ssp. *fulgida* or ssp. *sangamonensis*. It is distinguished by its glandular pubescence. Michaux's var. *virens* is nearly identical to ssp. *pilosa*.

Levin has annotated specimens from Fayette, Ford, Hardin, Johnson, and Montgomery counties as hybrids between ssp. *pilosa* and ssp. *fulgida*.

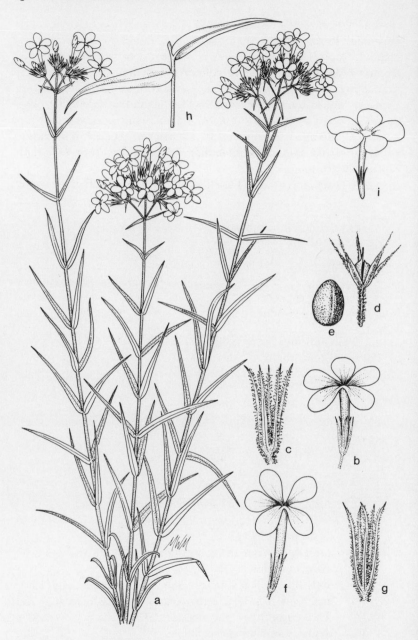

68. *Phlox pilosa* (Downy Phlox). *a.* Habit, × ½. *b.* Flower, × 1½. *c.* Calyx, × 3½. *d.* Fruit, × 5. *e.* Seed, × 5. ssp. *fulgida* (Downy Phlox). *f.* Flower, × 1½. *g.* Calyx, × 3½. ssp. *sangamonensis* (Sangamon Phlox). *h.* Leaves, × ½. *i.* Flower, × 1.

The flowers of ssp. *pilosa,* which show variation in color, open from May to August.

4b. Phlox pilosa L. ssp. **fulgida** (Wherry) Wherry, Baileya 4:97. 1956. *Fig. 68f, g.*
Phlox pilosa L. f. *albiflora* MacM. Metasp. Minn. Valley 432. 1892.
Phlox pilosa L. var. *fulgida* Wherry, Bartonia 12:47. 1931.
Phlox pilosa L. var. *fulgida* Wherry f. *albiflora* (MacM.) Standley, Rhodora 34:176. 1932.

Stems, leaves, and calyx pubescent with eglandular hairs.

COMMON NAME: Downy Phlox.
HABITAT: Prairies.
RANGE: Wisconsin to Saskatchewan, south to Kansas and Illinois.
ILLINOIS DISTRIBUTION: Occasional throughout the state.
There is considerable intergradation in Illinois between ssp. *fulgida* and ssp. *pilosa.* Levin has even found specimens he believes to be hybrids between these two subspecies.
White-flowered forms of ssp. *fulgida* are infrequently found in Illinois.
The flowers of ssp. *fulgida* bloom from May to August.

4c. Phlox pilosa L. ssp. **sangamonensis** Levin & Smith, Brittonia 17:264. 1965. *Fig. 68h, i.*
Phlox divaricata L. X *Phlox glaberrima* L., Ahles in Jones & Fuller, Vasc. Pl. Ill. 389. 1955.

Stems, leaves, and calyx glabrous or nearly so.

COMMON NAME: Sangamon Phlox.
HABITAT: Roadsides, fields, woods.
RANGE: Central Illinois.
ILLINOIS DISTRIBUTION: Known from Champaign and Piatt counties.
This subspecies was first described by Ahles as a hybrid between *Phlox divaricata* and *P. glaberrima.* Levin and Smith (1965) detail the exact status of this taxon.
The flowers bloom from May to August.

5. **Phlox paniculata** L. Sp. Pl. 151. 1753. *Fig. 69.*
Phlox acuminata Pursh, Fl. Am. Sept. 2:730. 1814.
Phlox paniculata L. var. *acuminata* (Pursh) Chapm. Fl. S. U. S. 338. 1860.

Perennial herb; stems erect, simple or branched, green, glabrous or puberulent, to 2 m tall; leaves oblong to lance-ovate, acute to acuminate at the apex, cuneate at the base, or the uppermost leaves subcordate at the base, entire, glabrous or puberulent, the lateral veins prominent, to 15 cm long; flowers in pyramidal panicles, on short pedicels; calyx to 1 cm long, green, glabrous or puberulent, 5-lobed, the lobes subulate, about half as long as the calyx tube; corolla salverform, to 2.5 cm long, pink or purple, 5-lobed; stamens 5, included; capsule oval, to 1 cm long.

COMMON NAMES: Garden Phlox; Sweet William.

HABITAT: Rich woods.

RANGE: New York to Iowa, south to Arkansas and Georgia.

ILLINOIS DISTRIBUTION: Scattered in Illinois, except for the northwestern counties.

Although this handsome species is often found in gardens, it is also a native member of our flora. Some of the collections from Illinois are from garden escapes. This species is one of several plants in Illinois called sweet william.

Phlox paniculata is related to *P. maculata,* but differs in the prominent leaf venation and absence of purple markings on the stem.

The flowers, which are clustered in huge pyramidal panicles, bloom from July to September.

6. **Phlox maculata** L. Sp. Pl. 152. 1753.

Perennial herb; stems erect, simple or branched, speckled with purple, glabrous or pubescent, to nearly 1 m tall; lower leaves linear to lanceolate, acuminate at the apex, rounded at the base, the upper lanceolate to lance-ovate, acuminate at the apex, rounded or subcordate at the base, entire, glabrous or puberulent, the lateral veins not prominent, to 10 cm long; flowers borne in narrow conical to cylindrical panicles, on short pedicels; calyx to 7 mm long, green, glabrous or puberulent, 5-lobed, the lobes lanceolate, acute to acu-

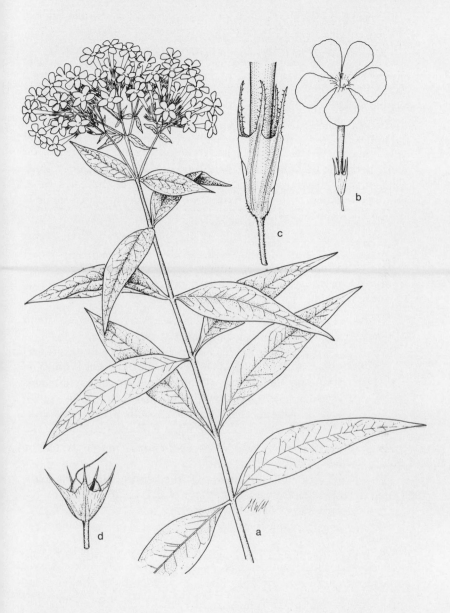

69. *Phlox paniculata* (Garden Phlox; Sweet William). *a.* Habit, ×½. *b.* Flower, ×1½. *c.* Calyx, ×5 *d.* Fruit, ×2½.

minate, about one-fourth as long as the calyx tube; corolla salver-form, to 2.5 cm long, pink or purple, 5-lobed; stamens 5, included; capsule oval, to 1 cm long.

This is another plant sometimes called sweet william. The lack of prominent leaf venation distinguishes it from *P. paniculata*, while the presence of purple speckles on the stems distinguishes it from *P. glaberrima* ssp. *interior* and *P. paniculata*.

Two subspecies occur in Illinois. They maybe differentiated by the following key:

KEY TO THE SUBSPECIES OF P. maculata IN ILLINOIS

1. Inflorescence narrow-conical; leaves lance-ovate _____
_____ 6. *P. maculata* ssp. *maculata*
1. Inflorescence cylindrical; leaves lanceolate _____
_____ 6b. *P. maculata* ssp. *pyramidalis*

6a. Phlox maculata L. ssp. **maculata** *Fig. 70a, b.*
Phlox suaveolens Ait. Hort. Kew. 1:206. 1789.
Phlox maculata L. var. *purpurea* Michx. Fl. Bor. Am. 1:143. 1803.

Inflorescence narrow-conical; leaves lance-ovate.

COMMON NAME: Speckled Phlox; Wild Sweet William.
HABITAT: Wet prairies; near bogs and swamps; in moist woodlands and meadows; along wooded streams.
RANGE: Quebec to Minnesota, south to Missouri and North Carolina.
ILLINOIS DISTRIBUTION: Occasional in the northern half of the state.
Levin (1963) indicates that the internodes of ssp. *maculata* are longer than those of ssp. *pyramidalis*.
This subspecies flowers during the latter half of May, usually about two weeks before ssp. *pyramidalis*.

6b. Phlox maculata L. ssp. **pyramidalis** (J. E. Smith) Wherry, Castanea 16:100. 1951. *Fig. 70c, d.*
Phlox pyramidalis J. E. Smith, Exot. Bot. 2:55. 1804.
Inflorescence cylindrical; leaves lanceolate.

70. *Phlox maculata* (Speckled Phlox; Wild Sweet William). *a*. Habit, × ½. *b*. Flower, × 1½. ssp. *pyramidalis* (Wild Sweet William). *c*. Habit, × ½. *d*. Flower, × 1½.

COMMON NAME: Wild Sweet William.

HABITAT: Somewhat drier than that of ssp. *maculata*.

RANGE: Mostly Appalachian, extending into Missouri and Illinois.

ILLINOIS DISTRIBUTION: Occasional in the northern half of the state.

Levin (1963) suggests that *P. maculata* ssp. *pyramidalis* is a hybrid derivative of *P. maculata* ssp. *maculata* and *P. glaberrima* ssp. *interior*.

This subspecies flowers during the first part of June, about two weeks after ssp. *maculata* flowers.

7. **Phlox glaberrima** L. ssp. **interior** (Wherry) Wherry, Baileya 4:98. 1956. *Fig. 71.*

Phlox glaberrima L. var. *interior* Wherry, Bartonia 14:19. 1932.

Perennial herb; stems erect, simple, glabrous, to 85 cm tall; leaves linear-lanceolate to lanceolate, acuminate at the apex, cuneate to the base, entire, glabrous, the lateral veins not prominent, to 10 cm long; flowers borne in a flat-topped cyme, on short pedicels; calyx to 7.5 mm long, green, glabrous, 5-lobed, the lobes subulate-lanceolate, less than half as long as the calyx tube; corolla salver-form, to 2.5 cm long, pink, 5-lobed; stamens 5, included; capsule oval, to 5 mm long.

COMMON NAME: Smooth Phlox.

HABITAT: Woods, prairies.

RANGE: Ohio to southern Wisconsin, south to Missouri and Kentucky.

ILLINOIS DISTRIBUTION: Occasional throughout the state.

Our specimens fall under ssp. *interior*, a midwestern taxon with the calyx shorter than the typical subspecies. This is a handsome species with showy rose-pink flowers.

Almost all parts of the plant are glabrous.

The flowers bloom from May to August.

8. **Phlox carolina** L. ssp. **angusta** Wherry, Baileya 4:98. 1956. *Fig. 72.*

Perennial herb; stems erect, simple or branched, green, usually glabrous, to 70 cm tall; leaves linear to lanceolate, acute to acumi-

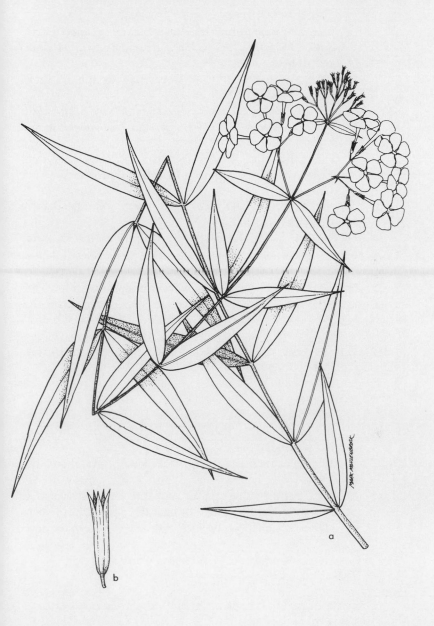

71. *Phlox glaberrima* ssp. *interior* (Smooth Phlox). *a.* Habit, × ½. *b.* Calyx, × 5.

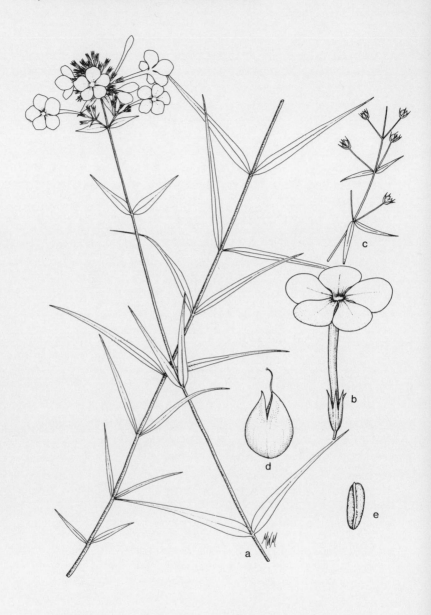

72. *Phlox carolina* ssp. *angusta* (Carolina Phlox). *a.* Habit, × ½. *b.* Flower, × 1½. *c.* Fruiting branchlet, × ½. *d.* Fruit, × 5. *e.* Seed, × 5.

nate at the apex, cuneate to somewhat rounded at the base, entire, usually glabrous, up to 10 cm long, up to 2 cm broad, the lateral veins not prominent; inflorescence cymose, usually about as broad as long, the flowers short-pedicellate; calyx 9–12 mm long, green, glabrous, 5-lobed, the lobes subulate-lanceolate, less than half as long as the calyx tube; corolla salverform, to 2.5 cm long, pink, 5-lobed; stamens 5, included; capsule oval, to 5 mm long.

COMMON NAME: Carolina Phlox.

HABITAT: Open areas (in Illinois).

RANGE: Pennsylvania to Missouri, south to Mississippi and Florida.

ILLINOIS DISTRIBUTION: Very rare and perhaps no longer existing in the state.

Wherry (1955) reported a specimen of *Phlox carolina* from Jefferson County. Efforts by Illinois botanists to relocate this species in Jefferson County in recent years have been unsuccessful. Levin has annotated a specimen from Macoupin County as *P. carolina* ssp. *angusta*. Some Illinois botanists have suggested that this specimen may actually be *P. glaberrima* ssp. *interior*.

The flowers bloom in late May and June.

6. *Microsteris* Greene–Microsteris

Annual herbs; leaves simple, entire, the lower leaves opposite, the upper leaves alternate; flowers perfect, actinomorphic, solitary or paired in the uppermost axils; calyx narrowly tubular to campanulate, 5-lobed; corolla salverform, 5-lobed; stamens 5, included; ovary superior, 3-locular; fruit a capsule with few seeds.

Microsteris is a genus of about six species native to the western United States.

Only the following species occurs in Illinois.

1. **Microsteris gracilis** (Dougl.) Greene, Pittonia 3:300. 1898.
   Fig. 73.
   Collomia gracilis Dougl. ex Benth. Bot. Reg. t. 1622.

Annual herb; stems much branched, pubescent, ascending to 50 cm; lower leaves opposite, entire, spatulate, obtuse at the apex, tapering to the base, pubescent, often smaller than the upper leaves; upper leaves alternate, entire, linear to lanceolate, obtuse to subacute at the apex, cuneate to the base, pubescent, up to 2 cm

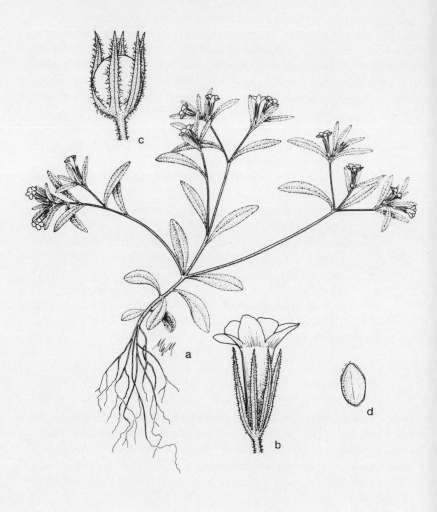

73. *Microsteris gracilis* (Microsteris). *a*. Habit, ×1. *b*. Flower, ×5. *c*. Fruit, ×5. *d*. Seed, ×5.

long, up to 5 mm broad; flower solitary in the axils of the uppermost leaves, very short-pedicellate; calyx 5-lobed, united at the base, pubescent, 4–7 mm long, the lobes linear, much longer than the tube; corolla 5-lobed, 6–9 mm long, narrowly tubular below, the lobes spreading, much shorter than the tube, the tube yellow, the lobes purplish; stamens 5, included; capsule oblongoid, about twice as long as the calyx tube.

COMMON NAME: Microsteris.

HABITAT: Roadside (in Illinois).

RANGE: Western United States; rarely adventive east of the Mississippi River.

ILLINOIS DISTRIBUTION: Known only from Macon Co.: Roadside, 2½ miles west of Niantic, May 16, 1957, *G. S. Winterringer s.n.*

This species is widespread in the western United States.

Its only Illinois collection was undoubtedly of an adventive plant.

This species flowers during late spring and most of the summer.

Order Campanulales

The Campanulales in Illinois consist only of the family Campanulaceae. The characters of the order are the same as the characters of the family.

CAMPANULACEAE–BELLFLOWER FAMILY

Annual or perennial herbs (in Illinois), rarely woody; latex sometimes present; leaves alternate, simple, without stipules; flowers perfect, actinomorphic or zygomorphic, borne in cymes, racemes, or panicles, or solitary; calyx 5-lobed; corolla campanulate or rotate or tubular, sometimes split down one side, 5-lobed; stamens 5, the filaments sometimes united; disk usually present; pistil 1, the ovary inferior, 2- or 5-locular, with numerous ovules borne on axile placentae, the style 1, the stigmas 2–5; fruit usually a capsule.

Although *Lobelia* and related genera have often been segregated into the Lobeliaceae, I am following the more recent conclusions of botanists that the Lobeliaceae and Campanulaceae should be combined. While the symmetry of the flowers of *Campanula* and *Lobelia* are clearly different, other characteristics of these genera are very similar. These similarities include presence of latex, alternate leaves, five stamens free from the corolla tube, an inferior ovary, and a capsular fruit.

Thus combined, the Campanulaceae have about sixty genera and over fifteen hundred species found in most parts of the World.

KEY TO THE GENERA OF Campanulaceae IN ILLINOIS

1. Flowers actinomorphic; corolla tube not split down one side _____ 2
1. Flowers zygomorphic; corolla tube split down one side ___ 3. *Lobelia*
    2. Leaves sessile, often clasping; flower solitary and sessile in the axils _____ 1. *Triodanis*
    2. Leaves petiolate, or at least not clasping at the base; flowers in spikes or racemes or, if solitary, then pedicellate _____ 2. *Campanula*

1. *Triodanis* Raf.–Venus' Looking-glass

Annual herbs; leaves alternate, simple, some or all the leaves usually clasping; flowers perfect, actinomorphic, solitary in the axils of leaflike bracts, the earliest flowers often small and cleistogamous; calyx tubular, (3-) 5-lobed; corolla rotate, 5-lobed; stamens 5, free;

ovary inferior, 3-locular, the stigma 3-lobed; capsule 3-locular, dehiscing by 3 lateral valves or pores, with numerous seeds.

Although species in this genus have often been placed in *Specularia*, McVaugh (1945) has given reasons for accepting *Triodanis* as the correct name. There are about a dozen species native to the northern hemisphere.

The lowermost flowers on any plant are cleistogamous, that is, without any corolla or only with a rudimentary corolla.

KEY TO THE SPECIES OF Triodanis IN ILLINOIS

1. Bracts suborbicular to ovate; seeds up to 0.7 mm long _____
_____ 1. *T. perfoliata*
1. Bracts linear to lanceolate; seeds 0.7–1.0 mm long __ 2. *T. leptocarpa*

1. **Triodanis perfoliata** (L.) Nieuwl. Am. Midl. Nat. 3:192. 1914.
   Campanula perfoliata L. Sp. Pl. 169. 1753.
   Campanula amplexicaulis Michx. Fl. Bor. Am. 1:108. 1803.
   Specularia perfoliata (L.) A. DC. Mon. Campan. 351. 1830.

Annual; stems erect, simple or branched, pilose, hispid or nearly glabrous, to 75 cm tall; leaves and bracts ovate to suborbicular, obtuse to acute at the apex, cordate at the sessile or clasping base, crenate, hirsutulous to glabrous, to 2.5 cm long, some of the basal leaves sometimes on distinct petioles; cleistogamous flowers borne at most of the nodes; calyx of cleistogamous flowers 3- to 4-parted, the lobes lanceolate, ovate, or subulate; calyx of petaliferous flowers 5-parted, the lobes lanceolate to subulate-lanceolate; corolla of petaliferous flowers bluish purple or rarely white; capsule ellipsoid to oblongoid, to 2 cm long, the pores near the apex or midway between the base and the apex; seeds lenticular, ellipsoid, 0.5–0.7 mm long.

There are two varieties of *T. perfoliata* in Illinois.

KEY TO THE VARIETIES OF T. perfoliata IN ILLINOIS

1. Capsule with pores midway between base and apex _____
_____ 1a. *T. perfoliata* var. *perfoliata*
1. Capsule with pores near apex 1b. *T. perfoliata* var. *biflora*

1a. **Triodenis perfoliata** (L.) Nieuwl. var. **perfoliata** *Fig. 74.*
Triodanis perfoliata (L.) Nieuwl. f. *alba* Voigt in Mohlenbr. & Voigt, Fl. S. Ill. 325. 1959.

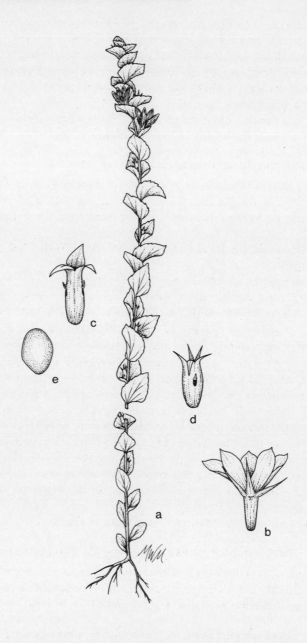

74. Triodanis perfoliata (Venus' Looking-glass). *a.* Habit, ×½. *b.* Flower, ×2. *c.* Cleistogamous flower, ×2½. *d.* Fruit, ×2½. *e.* Seed, ×25.

Specularia perfoliata (L.) A. DC. f. *alba* (Voigt) Steyerm. Rhodora 62:131. 1960.

Capsule with pores midway between base and apex.

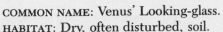 COMMON NAME: Venus' Looking-glass.
HABITAT: Dry, often disturbed, soil.
RANGE: Quebec to British Columbia, south to Montana, Texas, and Florida; Mexico.
ILLINOIS DISTRIBUTION: Common; probably in every county.
This Venus' looking-glass is common in dry, often weedy, habitats throughout the state. The pores of the capsule midway between the base and the apex distinguish this variety from var. *biflora*.

A white-flowered form was originally described from southern Illinois.

The flowers bloom from late April to August.

1b. Triodanis perfoliata (L.) Nieuwl. var. **biflora** (Ruiz & Pavon) Bradley, Brittonia 27:114. 1975 *Fig. 75.*
Campanula biflora Ruiz & Pavon, Fl. Peruv. 2:55. 1799.
Specularia biflora (Ruiz & Pavon) Fisch. & Mey. Ind. Sem. Hort. Petrop. 1:17. 1836.
Triodanis biflora (Ruiz & Pavon) Greene, Man. Bot. San Francisco Bay 230. 1894.

Capsule with pores near apex.

COMMON NAME: Venus' Looking-glass.
HABITAT: Dry, often disturbed, soil.
RANGE: Virginia to Kansas, south to Texas and Florida; Oregon; Mexico.
ILLINOIS DISTRIBUTION: Confined to the southern one-sixth of the state.
The only sure way to distinguish this variety from the very similar typical *T. perfoliata* is by the nearly apical pores of the capsule. In general, var. *biflora* has fewer petaliferous flowers, slightly narrower leaves and bracts that often are not clasping, and less pubescence. None of these last characters are always reliable, however.

Several botanists consider this taxon to be a distinct species.

The flowers bloom from late April to June.

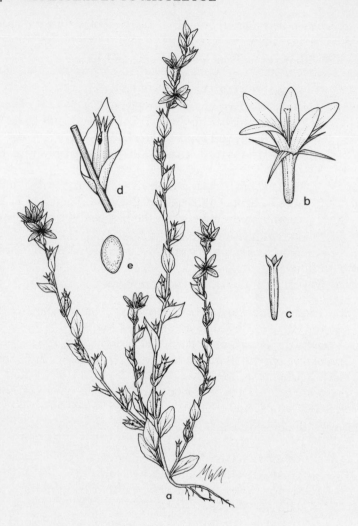

75. *Triodanis perfoliata* var. *biflora* (Venus' Looking-glass). *a.* Habit, × ½. *b.* Flower, × 2. *c.* Cleistogamous flower, × 3. *d.* Fruit, × 2. *e.* Seed, × 10.

2. Triodenis leptocarpa (Nutt.) Nieuwl. Am. Midl. Nat. 3:192. 1914. *Fig. 76.*
Campylocera leptocarpa Nutt. Trans. Am. Phil. Soc. 8:247. 1843.
Specularia leptocarpa (Nutt.) Gray, Proc. Am. Acad. 11:82. 1876.

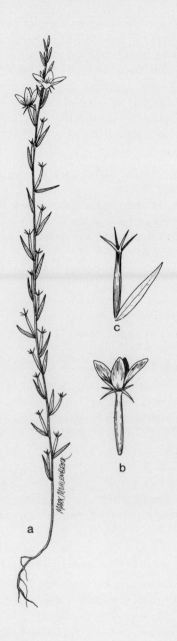

76. *Triodanis leptocarpa* (Slender-leaved Venus' Looking-glass). *a*. Habit, × ½. *b*. Flower, × 1. *c*. Fruit, × 1.

Annual; stems erect, simple or branched, hirsutulous, to 75 cm tall; leaves and bracts linear to lanceolate, obtuse to acute at the apex, cuneate to the sessile base, entire, hirsutulous to glabrous, to 2.5 cm long; cleistogamous flowers only at the lowest floral nodes; calyx of cleistogamous flowers 3-parted, the lobes subulate; calyx of petaliferous flowers 5-parted, the lobes subulate; corolla of petaliferous flowers bluish-purple; capsule linear-cylindric, to 2 cm long, the pores near the apex, sometimes slitlike.

COMMON NAME: Slender-leaved Venus' Looking-glass.
HABITAT: Fields.
RANGE: Indiana and Minnesota to Montana, south to Oklahoma and Arkansas.
ILLINOIS DISTRIBUTION: Known only from the northern half of the state.
This rare western species is known from a few locations in northern counties of the state.
The narrow leaves and bracts easily distinguish it from S. *perfoliata*.
The flowers appear from May to July.

2. *Campanula* L.–Bellflower

Annual or perennial herbs; leaves alternate or basal, simple, not clasping; flowers perfect, actinomorphic, solitary or in racemes or panicles; calyx 5-lobed; corolla campanulate or rotate, 5-lobed; stamens 5, free; ovary inferior, 3- to 5-locular, the stigma 3-to 5-lobed; capsule 3- to 5-locular, dehiscing by 3–5 lateral valves or pores, with numerous seeds.

There are about two hundred fifty species of *Campanula*, most of them native in the Northern Hemisphere, particularly in Europe. Because of their attractive flowers, many are grown as garden ornamentals.

KEY TO THE SPECIES OF Campanula IN ILLINOIS

1. Cauline leaves linear to narrowly lanceolate (ovate, cordate basal leaves may be present at base of plant in *C. rotundifolia*) _____ 2
1. Cauline leaves oblong-lanceolate to ovate_____ 4
    2. Stems glabrous or rarely closely puberulent; corolla 1.5 cm long or longer _____ 1. *C. rotundifolia*
    2. Stems scabrous with retrorse hairs on the angles; corolla up to 1.2 cm long_____ 3

3. Corolla white, 5–8 mm long; capsule up to 2 mm long _____
_____ 2. *C. aparinoides*
3. Corolla bluish, 10–12 mm long; capsule 3–5 mm long _____
_____ 3. *C. uliginosa*
4. Corolla campanulate; capsule with pores nearly basal _____ 5
4. Corolla rotate; capsule with pores nearly apical_____
_____ 6. *C. americana*
5. Flowers short-pedicellate, drooping, borne in elongated, 1-sided racemes; lobes of calyx linear; capsule globose, nodding _____
_____ 4. *C. rapunculoides*
5. Flowers sessile, erect, borne in glomerules; lobes of calyx lanceolate;
capsule ovoid, erect _____ 5. *C. glomerata*

1. **Campanula rotundifolia** L. Sp. Pl. 163. 1753.
Perennial herb from slender rootstocks; stems erect or spreading,
often branched from the base, glabrous or rarely closely pubescent,
to 45 cm tall; basal leaves orbicular to ovate, acute at the apex,
cordate at the base, dentate to entire, glabrous or rarely closely
pubescent, to 2.5 cm long, slenderly petiolate, usually absent at
flowering time; cauline leaves linear to narrowly lanceolate, acute
at the apex, cuneate to the nearly sessile base, entire or nearly so,
glabrous or rarely closely puberulent; flowers solitary or in racemes, borne on divergent or pendulous pedicels; calyx deeply 5-
parted, the lobes subulate, 5–15 mm long, much longer than the
calyx tube; corolla campanulate, blue, 1.5–2.5 cm long, shallowly
5-lobed; capsule ovoid, pendulous, strongly ribbed, opening by
basal pores or slits.
Two varieties are recognizable in Illinois, distinguished as fol-
lows:

KEY TO THE VARIETIES OF C. rotundifolia IN ILLINOIS

1. Stems and cauline leaves glabrous _____
_____ 1a. *C. rotundifolia* var. *rotundifolia*
1. Stems and cauline leaves closely puberulent _____
_____ 1b. *C. rotundifolia* var. *velutina*

1a. **Campanula rotundifolia** L. var. **rotundifolia** *Fig. 77.*
Campanula intercedens Witasek, Abh. Zool. Bot. Ges. Wien
1(3):43. 1802.
Campanula rotundifolia L. var. *intercedens* (Witasek) Farw. Pa-
pers Mich. Acad. Sci. 3:105. 1924.
Stems and cauline leaves glabrous.

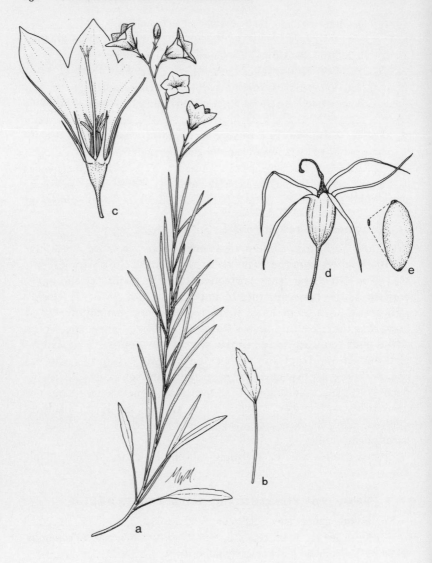

77. *Campanula rotundifolia* (Bellflower; Harebell). *a.* Habit, ×½. *b.* Basal leaf, ×½. *c.* Flower (partly cut away), ×2½. *d.* Fruit, ×2. *e.* Seed, ×20.

COMMON NAMES: Bellflower; Harebell.
HABITAT: Woods, hill prairies, sandstone cliffs.
RANGE: Nova Scotia to British Columbia, south to California, Arizona, Texas, and West Virginia; Europe; Asia.
ILLINOIS DISTRIBUTION: Occasional in the northern half of the state; also Jackson and Williamson counties. The smooth variety is the common one in Illinois.

In the northern half of the state, *C. rotundifolia* may be found in sandy black oak woods, on hill prairies, or on shaded rocky cliffs. As its two stations in southern Illinois, it grows on moist sandstone ledges.

The specific epithet refers to the suborbicular basal leaves that usually are absent at flowering time.

This variety flowers from June to September.

1b. Campanula rotundifolia L. var. **velutina** A. DC. Mon. Campan. 1830. *Not illustrated.*

Stem and leaves closely puberulent.

COMMON NAME: Bellflower.
HABITAT: Crevices of cliffs.
RANGE: Similar to that of var. *rotundifolia*.
ILLINOIS DISTRIBUTION: Known only from Jo Daviess County.

The pubescent var. *velutina* is apparently extremely rare in Illinois.

It flowers from July to September.

2. Campanula aparinoides Pursh, Fl. Am. Sept. 159. 1814. *Fig. 78.*

Perennial herb from slender rootstocks; stems weak and reclining, branched, retrorsely hispid, to 50 cm long; leaves linear-lanceolate to lanceolate, acute at the apex, cuneate to the sessile base, crenulate to denticulate, pubescent, to 4.5 cm long; flowers borne in leafy panicles from filiform, divergent pedicels; calyx 1.5–4.0 mm long, 5-parted, the lobes lanceolate, about as long as the calyx tube; corolla campanulate, white, 5–8 mm long, deeply 5-lobed; capsule subglobose, ascending, ribbed, to 2 mm long, opening by basal pores.

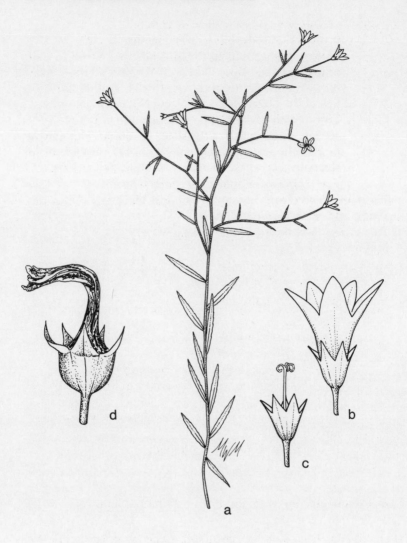

78. *Campanula aparinoides* (Marsh Bellflower). *a.* Habit, ×½. *b.* Flower, ×3½. *c.* Flower (corolla removed), ×3½. *d.* Developing fruit, ×5.

COMMON NAME: Marsh Bellflower.

HABITAT: Marshes and bogs.

RANGE: Maine to Minnesota, south to Colorado, Missouri, and Georgia.

ILLINOIS DISTRIBUTION: Occasional in the northern half of the state.

This and the following species are closely related and are actually treated as one species by Shetler (1963). Both have the appearance of a *Galium* because of the sprawling growth form and narrow leaves.

The differences exhibited by *C. aparinoides* are the smaller white flowers and the very tiny capsules.

Swink and Wilhelm (1979) record this species from marshes, calcareous fens, and bogs.

It flowers from June to August.

3. **Campanula uliginosa** Rydb. in Britt. Man. Fl. N. States 885. 1901. *Fig. 79.*

Campanula aparinoides Pursh var. *grandiflora* Holz. Bull. Geol. & Nat. Hist. Surv. Minn. 9:566. 1896.

Campanula aparinoides Pursh var. *uliginosa* (Rydb.) Gl. Phytologia 4:25. 1952.

Perennial herb from slender rootstocks; stems weak and reclining, branched, retrorsely hispid, to 50 cm long; leaves linear to linear-lanceolate, acuminate at the apex, cuneate to the sessile base, denticulate, pubescent, to 5 cm long; flowers borne on nearly bractless, ascending pedicels; calyx 3–7 mm long, 5-parted, the lobes usually a little longer than the calyx tube; corolla campanulate, bluish, 10–12 mm long, 5-lobed, the lobes about equaling the corolla tube; capsule subglobose, ascending, ribbed, 3–5 mm long, opening by basal pores.

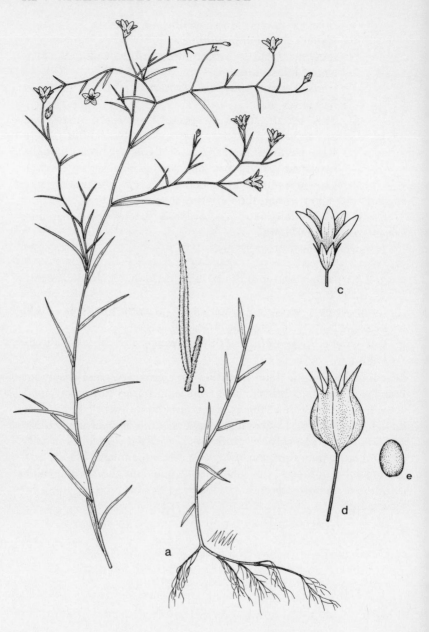

79. *Campanula uliginosa* (Marsh Bellflower). *a.* Habit, × ½. *b.* Leaf, × 1. *c.* Flower, × 2. *d.* Fruit, × 5. *e.* Seed, × 15.

COMMON NAME: Marsh Bellflower.

HABITAT: Marshes and bogs.

RANGE: Quebec to Saskatchewan, south to Nebraska, Illinois, and New York.

ILLINOIS DISTRIBUTION: Occasional in the northern half of the state.

Shetler (1963) considers this plant synonymous with *C. aparinoides*, and Gleason (1952) places it as a variety of *C. aparinoides*, calling it var. *uliginosa*. There seems to be as many good reasons to accept it as a distinct species, however.

This species flowers from June to August.

4. Campanula rapunculoides L. Sp. Pl. 165. 1753. *Fig. 80.*

Perennial from slender rootstocks; stems erect, simple or branched, glabrous or finely pubescent, to 75 cm tall; leaves ovate to ovate-lanceolate, acute to acuminate at the apex, cordate or rounded at the base, hispidulous on the lower surface, to 15 cm long, crenate or denticulate, the lowermost petiolate, the upper sessile; flowers in 1-sided racemes, on pendulous pedicels, with slender bracts; calyx deeply 5-lobed, the lobes linear, pubescent; corolla campanulate, blue, to 3 cm long, 5-lobed; capsule globose, pendulous, pubescent, to nearly 1 cm in diameter, opening by basal pores.

COMMON NAME: European Bellflower.

HABITAT: Roadsides and other waste areas.

RANGE: Native of Europe; adventive in northeastern North America.

ILLINOIS DISTRIBUTION: Apparently confined to the northern half of the state.

The European bellflower is sometimes planted as a garden ornamental. It spreads rapidly from underground rootstocks.

This species is similar to another introduced species, *C. glomerata*, differing by its 1-sided inflorescence with pedicellate, drooping flowers, its linear calyx lobes, and its drooping, globose capsules.

The pretty bell-shaped flowers bloom from July to September.

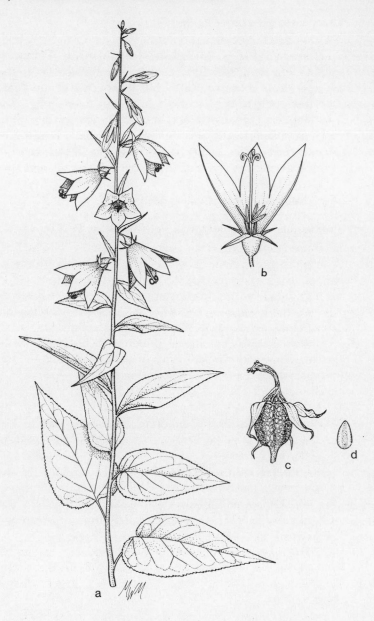

80. Campanula rapunculoides (European Bellflower). *a.* Habit, × ½. *b.* Flower (partly cut away), × 1. *c.* Developing fruit, × 1½. *d.* Seed, × 5.

5. **Campanula glomerata** L. Sp. Pl. 166. 1753. *Fig. 81.*

Perennial from slender rootstocks; stems erect, simple or branched, usually pubescent, up to 70 cm tall; leaves pubescent, the lowermost petiolate, oblong to oblong-lanceolate, usually tapering to the base, the uppermost ovate to ovate-lanceolate, acute at the apex, tapering to the sessile or subclasping base, up to 10 cm long, denticulate to crenulate; flowers erect, sessile, in glomerules, with leafy bracts; calyx deeply 5-lobed, the lobes lanceolate, pubescent; corolla campanulate, blue, to 25 (–30) cm long, 5-lobed; capsule ovoid, erect, pubescent, to 7 mm high, opening by basal pores.

 COMMON NAME: Clustered Bellflower.

HABITAT: Waste ground.

RANGE: Native of Europe and Asia; infrequently adventive in the eastern United States.

ILLINOIS DISTRIBUTION: Known only from Wabash County.

This species, sometimes grown as a garden ornamental, has been found a single time in Illinois as an adventive. It flowers from June through August.

6. **Campanula americana** L. Sp. Pl. 164. 1753. *Fig. 82.*

Campanula illinoensis Fresn. in DC. Prodr. 7:478. 1838.

Campanula americana L. var. *illinoensis* (Fresn.) Farw. Rep. Mich. Acad. Sci. 20:191. 1918.

Annual herbs; stems stout, erect, usually simple, pubescent, to 2 m tall; leaves lanceolate to ovate, acuminate at the apex, cuneate to the base, serrate, pubescent, short-petiolate, or the uppermost sessile, to 15 cm long; flowers in racemes, on short pedicels from the axils of bracts, the lowermost bracts foliaceous; calyx deeply 5-lobed, the lobes linear-subulate; corolla rotate, blue, to 2.5 cm across, deeply 5-lobed; style declined but turned upward at the tip; capsule cylindric to turbinate, ribbed, glabrous, to 1 cm long, opening by subapical pores.

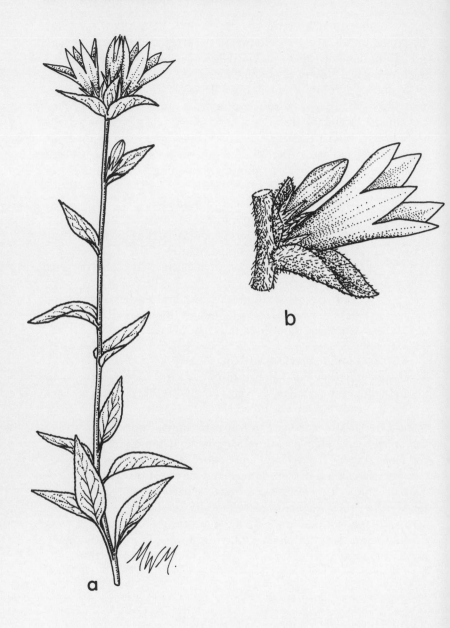

81. *Campanula glomerata* (Clustered Bellflower). *a.* Habit, × ½. *b.* Flower and bud, × 1.

82. *Campanula americana* (Tall Bellflower; American Bellflower). *a.* Habit, × ½. *b.* Flower, × 1. *c.* Fruit, × 2. *d.* Seed, × 10.

COMMON NAMES: Tall Bellflower; American Bellflower.
HABITAT: Woods.
RANGE: Ontario to Minnesota, south to Oklahoma, Alabama, and Florida.
ILLINOIS DISTRIBUTION: Common; known from almost every county.
This is one of the prettiest of the midsummer wild flowers in Illinois. The rotate corolla sets this species well apart from the other Campanulas in Illinois that have campanulate corollas. The declined style, which is upturned at the tip, is also distinctive.

Variety *illinoensis* (Fresn.) Farw., with broader leaves, is not worthy of recognition.

The flowers bloom from June to October.

3. *Lobelia* L. –Lobelia

Herbs (in Illinois) or shrubs; leaves alternate or basal, simple; flowers perfect, zygomorphic, borne in spikes, racemes, or panicles; calyx tubular, 5-lobed; corolla tubular, split down one side, 2-lipped; stamens 5, the anthers united in a ring around the style, at least 2 of the anthers bearded at the tip; pistil 1, the ovary inferior, 2-locular; stigma 2-lobed; capsule 2-valved, dehiscing at the top, with numerous seeds.

Lobelia is a genus of about two hundred fifty species found in most parts of the World. It is sometimes placed in its own family. The corolla lobes on either side are usually recurved and turned away from the other three lobes.

KEY TO THE TAXA Lobelia IN ILLINOIS

1. Flowers 1.5 cm long or longer _____ 2
2. Flowers up to 1.5 cm long _____ 5
    2. Flowers red or deep rose (rarely white) _____ 3
    2. Flowers blue (rarely white) _____ 4
3. Flowers red; calyx glabrous or puberulent _____ 1. *L. cardinalis*
3. Flowers deep rose; calyx hirsute _____ 2. *L. X speciosa*
    4. Auricles at base of calyx 2–5 mm long; stems glabrous or sparsely hirsute; flowers 2.0–3.3 cm long _____ 3. *L. siphilitica*
    4. Auricles at base of calyx less than 2 mm long; stems densely puberulent throughout; flowers 1.5–2.5 cm long _____ 4. *L. puberula*
5. Cauline leaves linear to narrowly lanceolate; lower lip of corolla ·glabrous _____ 5. *L. kalmii*

5. Cauline leaves oblong, lanceolate, or obovate _____ 6
  6. Lower part of stems villous or hirsute; capsules inflated, completely enclosed by the calyx_____ 6. *L. inflata*
  6. Stems glabrous or short-pubescent; capsules not inflated, partly exserted from the calyx _____ 7. *L. spicata*

1. Lobelia cardinalis L. Sp. Pl. 930. 1753. *Fig. 83.*

Perennial with basal offshoots; stems erect, simple, glabrous or nearly so, to 1.5 m tall; leaves lanceolate to ovate-lanceolate, acute or acuminate at the apex, cuneate to the base, serrulate, glabrous or hirtellous, to 15 cm long, the lowermost petiolate, the upper sessile; flowers in racemes, borne on pedicels shorter than the glandular, leafy bracts; calyx deeply 5-lobed, the lobes linear, glabrous or pubescent; corolla scarlet (rarely white), 2.5–4.0 cm long, 2-lipped, 5-lobed; filament tube exserted; capsule glabrous; seeds wrinkled.

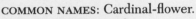

COMMON NAMES: Cardinal-flower.
HABITAT: Wet ground.
RANGE: New Brunswick to Ontario, south to Texas and Florida.
ILLINOIS DISTRIBUTION: Occasional to common throughout the state.
This is one of the most beautiful wild flowers in Illinois. The scarlet inflorescence, when in full bloom from July to September, is a sight to behold.
Although the hybrid formed between this species and *L. siphilitica* is similar in appearance, it can be distinguished by its rose-colored flowers and its hirtellous calyces. White-flowered specimens rarely occur.

2. Lobelia X speciosa Sweet, Brit. Flow. Gard. Ser. II, t. 174. 1833. *Not illustrated.*

Lobelia siphilitica L. var. *hybrida* Hack. Curtis' Bot. Mag. 64: tab. 3604. 1837.

Perennial with basal offshoots; stems erect, simple, usually puberulent, to 1 m tall; leaves lanceolate to ovate-lanceoate, acute or acuminate at the apex, cuneate to the base, serrulate, usually pubescent, to 15 cm long, the lowermost petiolate, the upper sessile; calyx deeply 5-lobed, the lobes linear-lanceolate, hirsute; corolla deep rose, 2.0–3.5 mm long, 2-lipped; capsule not seen.

83. *Lobelia cardinalis* (Cardinal-flower). *a.* Habit, × ½. *b.* Leaves, × ½. *c.* Flower, × 1. *d.* Developing fruit, × 1. *e.* Seed, × 10.

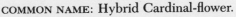

COMMON NAME: Hybrid Cardinal-flower.

HABITAT: Low areas.

RANGE: Sporadic in the eastern United States.

ILLINOIS DISTRIBUTION: Known only from Wabash County.

This remarkable plant is a hybrid between the brilliant red-flowered *L. cardinalis* and the equally handsome blue-flowered *L. siphilitica*. The flower color of the hybrid is rose. The calyx is hirsute, similar to that found in *L. siphilitica*.

Jacob Schneck first found this hybrid in Wabash County during the last part of the nineteenth century. It has not been found in Illinois since.

The flowers bloom during late summer.

3. **Lobelia siphilitica** L. Sp. Pl. 931. 1753. *Fig. 84.*

Lobelia siphilitica L. var. *ludoviciana* A. DC. in DC. Prodr. 7:377. 1839.

Perennial with basal offshoots; stems erect, simple, glabrous or puberulent, to 1 m tall; leaves lanceolate to oblong to ovate, acute or acuminate at the apex, cuneate to the base, serrulate to denticulate, glabrous or strigose, to 15 cm long, the lowermost petiolate, the upper sessile; flowers densely racemose, borne on short pedicels subtended by leafy bracts; calyx deeply 5-lobed, the lobes lanceolate, usually hirsute, with large reflexed auricles in the sinuses; corolla blue, except for the white base of the lower lip, 2.0–3.3 cm long, 5-lobed, the lobes about as long as the tube; capsule glabrous; seeds wrinkled.

COMMON NAME: Blue Cardinal-flower.

HABITAT: Wet ground.

RANGE: Maine to South Dakota, south to Texas and Alabama.

ILLINOIS DISTRIBUTION: Occasional to common throughout the state.

This beautiful species shows considerable variation in the degree of pubescence on the stems, leaves, and calyces.

McVaugh (1936) recognizes specimens with nearly glabrous leaves, stems, and calyces as var. *ludoviciana*, but I am unable to make such a distinction in the material available to me.

84. Lobelia siphilitica (Blue Cardinal-flower). *a.* Habit, ×½. *b.* Leaf, ×½. *c.* Flower, ×1½. *d.* Developing fruit, ×1. *e.* Seed, ×10.

Lobelia siphilitica can be distinguished from *L. puberula* by its larger flowers and its larger calyx auricles.

The flowers bloom from August to October.

4. Lobelia puberula Michx. Fl. Bor. Am. 2:152. 1803. *Fig. 85.*
Lobelia puberula Michx. var. *sumulans* Fern. Rhodora 49:184. 1947.

Perennial with basal offshoots; stems erect, simple, puberulent, to 1.5 m tall; leaves oblong to obovate, obtuse or subacute at the apex, cuneate to the base, denticulate to nearly entire, glabrous or puberulent, to 10 cm long; flowers racemose, often 1-sided, borne on short pedicels subtended by leafy bracts; calyx deeply 5-lobed, the lobes lanceolate, glabrous or puberulent, with tiny reflexed auricles in the sinuses, or auricles absent; corolla blue, except for the white base of the lower lip, 1.5–2.5 cm long, 5-lobed, the lobes usually not quite as long as the tube; capsule puberulent; seeds wrinkled.

COMMON NAME: Downy Lobelia.
HABITAT: Wet ground.
RANGE: New Jersey to Missouri, south to Texas and Florida.
ILLINOIS DISTRIBUTION: Occasional in the southern one-third of the state.

Fernald (1950) considers our specimens to belong to var. *simulans,* differing from typical var. *puberula* by its puberulent stems, its spreading leaves, its shorter calyx lobes, and its narrower bracts. I am unable to make adequate distinctions between the varieties enumerated by Fernald.

McVaugh (1936) cites a specimen from Henry County, a considerable distance north of the remainder of the Illinois range for *L. puberula.* I have not been able to verify the existence of the specimen on which the report was made.

The flowers bloom from August to October.

5. Lobelia kalmii L. Sp. Pl. 930. 1753. *Fig. 86.*

Perennial with basal offshoots; stems erect, branched, glabrous to pubescent, to 60 cm tall; basal leaves spatulate, obtuse at the apex, cuneate to the petiolate base, denticulate, pubescent, to 2.5 cm long; upper leaves linear to linear-lanceolate, acute or obtuse at the apex, cuneate to the sessile base, denticulate, pubescent to nearly glabrous, to 2 cm long; flowers in loose racemes, on slender pedi-

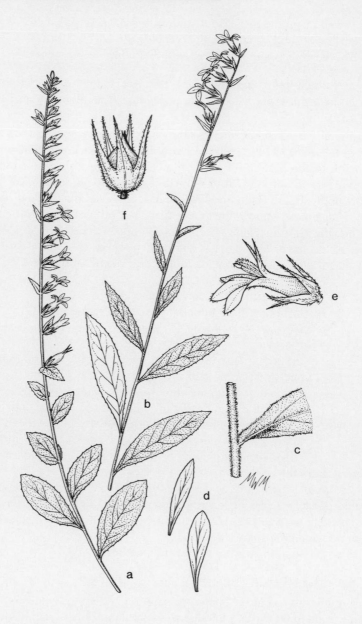

85. Lobelia puberula (Downy Lobelia). *a,b.* Habit, × ½. *c.* Leaf base, × 1½. *d.* Leaf variations, × ½. *e.* Flower, × 1½. *f.* Fruit, × 2.

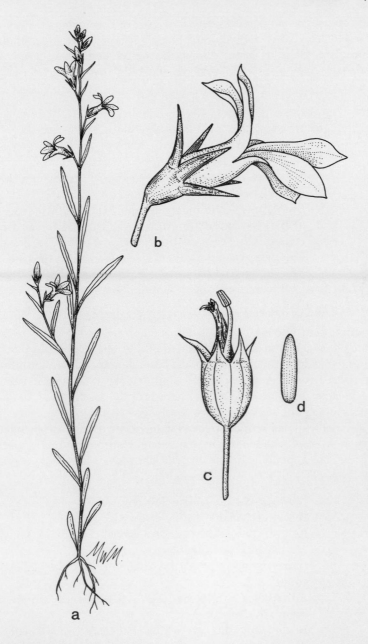

86. *Lobelia kalmii* (Kalm's Lobelia). *a.* Habit, ×½. *b.* Flower, ×3. *c.* Developing fruit, ×3½. *d.* Seed, ×15.

cels to 2.5 cm long, subtended by a pair of bracteoles midway on the pedicels; calyx campanulate, deeply 5-lobed, the lobes linear-lanceolate, glabrous or hirtellous, the sinuses not appendaged; corolla pale blue, 7–15 mm long; capsule subglobose, to 4 mm in diameter.

COMMON NAME: Kalm's Lobelia.
HABITAT: Springy areas and dunes.
RANGE: Newfoundland to Mackenzie, south to Colorado, northern Illinois, and New Jersey.
ILLINOIS DISTRIBUTION: Occasional in the northern one-third of the state.
Swink and Wilhelm (1979) report this northern species as being frequent in calcareous springy sites where it consistently occurs with *Parnassia glauca*.
The very narrow cauline leaves distinguish this species from *L. spicata*.
The flowers bloom from July to September.

6. Lobelia inflata L. Sp. Pl. 931. 1753. *Fig. 87.*

Annual from fibrous roots, stems erect, branched, villous or hirsute, to nearly 1 m tall; basal leaves oval to obovate, obtuse at the apex, cuneate to the short-petiolate base, dentate or serrate, pubescent, to 6 cm long; upper leaves obovate to ovate, acute to obtuse at the apex, cuneate or slightly rounded at the sessile base, dentate or serrate, pubescent, to 5 cm long; flowers in loose racemes, on short pedicels subtended by subulate bracts; calyx campanulate, deeply 5-lobed, the lobes linear-subulate, glabrous or nearly so, the sinuses not appendaged; corolla pale blue, 5–10 mm long; capsule globose, inflated, to 9 mm in diameter.

COMMON NAME: Indian Tobacco.
HABITAT: Woods, fields, disturbed areas.
RANGE: Labrador to Saskatchewan, south to Kansas, Mississippi, and Georgia.
ILLINOIS DISTRIBUTION: Common; probably in every county.
Although a native species, Indian tobacco commonly occurs in disturbed soil.
In the past, this species has been used medicinally in the treatment of laryngitis and asthma. The fruits are poisonous.
The flowers open from June to October.

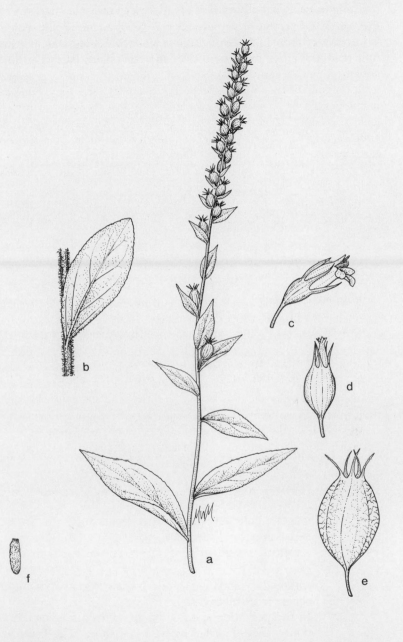

87. *Lobelia inflata* (Indian Tobacco). *a.* Habit, × ½. *b.* Leaf, × ½. *c.* Flower, × 2½. *d.* Developing fruit, × 2½. *e.* Fruit, × 2½. *f.* Seed, × 20.

7. **Lobelia spicata** Lam. Encycl. 3:587. 1789.

Perennial herb; stems erect, usually simple, glabrous to puberulent to hirtellous, to about 1 m tall; basal leaves oblanceolate to obovate, obtuse at the apex, cuneate to the short-petiolate base, dentate, crenate, or even entire, strigose to hirtellous, to 8 cm long; upper leaves oblong to lanceolate, obtuse to subacute at the apex, cuneate to the sessile base, dentate, crenate, or entire, strigose to hirtellous, spreading or ascending, to 6.5 cm long; flowers in dense or loose racemes, on short pedicels subtended by linear bracts; calyx campanulate, deeply 5-lobed, the lobes linear-subulate, the sinuses with reflexed auricles to 5 mm long; corolla pale blue to nearly white, 7–10 mm long; capsule globose.

Lobelia spicata is a variable species. Although four varieties have been attributed to Illinois, I am able to recognize only two of them.

KEY TO THE VARIETIES OF L. spicata IN ILLINOIS

1. Appendages between calyx lobes up to 1 mm long; leaves mostly spreading _____ 7a. *L. spicata* var. *spicata*
1. Appendages between calyx lobes 2–5 mm long; leaves mostly ascending _____ 7b. *L. spicata* var. *leptostachys*

7a. **Lobelia spicata** Lam. var. **spicata** *Fig. 88.*

Lobelia claytoniana Michx. Fl. Bor. Am. 2:153. 1803.

Lobelia spicata Lam var. *hirtella* Gray, Syn. Fl. N. Am. 2:6. 1878.

Lobelia spicata Lam var. *originalis* McVaugh, Rhodora 38:312. 1936.

Lobelia spicata Lam. var. *campanulata* McVaugh, Rhodora 38:316. 1936.

Appendages between calyx lobes up to 1 mm long; leaves mostly spreading.

88. *Lobelia spicata* (Spiked Lobelia). *a.* Habit, × ½. *b.* Flower, × 2½. *c.* Fruit, with corolla persisting, × 2½. *d.* Seed, × 20.

COMMON NAME: Spiked Lobelia.

HABITAT: Dry woods, prairies.

RANGE: Nova Scotia to Manitoba, south to Nebraska, Arkansas, and Georgia.

ILLINOIS DISTRIBUTION: Occasional throughout the state.

Considerable variation occurs in this variety, and many of the variants have been named. Var. *hirtella* has hirtellous stems and leaves, while var. *campanulata* has white anthers and dark purple-blue flowers, as well as a slightly campanulate calyx. To me, var. *hirtella* intergrades imperceptibly into typical var. *spicata*. Var. *campanulata* may merit varietal status, but I have seen too few specimens to evaluate the situation properly.

The flowers bloom from July to August.

7b. Lobelia speicata Lam. var. **leptostachys** (A. DC.) Mack. & Bush, Fl. Jackson Co., Mo. 183. 1902. *Fig. 89.*

Lobelia leptostachys A. DC. in DC. Prodr. 7:376. 1839.

Appendages between calyx lobes 2–5 mm long; leaves mostly ascending.

COMMON NAME: Spiked Lobelia.

HABITAT: Dry soil.

RANGE: West Virginia to eastern Kansas, south to eastern Texas and Alabama.

ILLINOIS DISTRIBUTION: Occasional throughout the state.

Although some botanists treat this taxon as a distinct species, it differs from *L. spicata* essentially only by its longer auricles between the calyx lobes.

The flowers bloom from June to August.

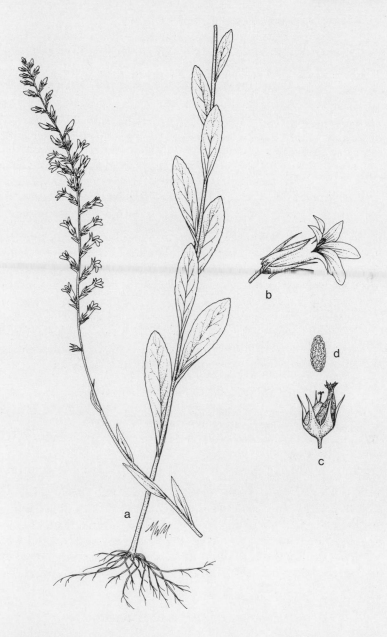

89. *Lobelia spicata* var. *leptostachys* (Spiked Lobelia). *a*. Habit, × ½. *b*. Flower, × 2½. *c*. Developing fruit, × 2½. *d*. Seed, × 20.

Order Santalales

The Santalales are represented in Illinois by three families—Celastraceae, Santalaceae, and Viscaceae. Members of the Celastraceae are completely self-sufficient and do not depend on other organisms for their support. The Santalaceae are partially parasitic on the roots of various woody species, while the Viscaceae are parasitic on the branches of various trees.

CELASTRACEAE—BITTERSWEET FAMILY

Woody shrubs, vines, or creepers; leaves alternate or opposite, simple, pinnately veined; flowers in axillary clusters or in terminal inflorescences or solitary, perfect or unisexual, actinomorphic; calyx 4- to 5-parted, very small; petals 4 or 5; free; stamens 4–5, attached to a disk; ovary superior, 2- to 5-locular, with two ovules per locule; fruit a capsule; seeds arillate.

This family is composed of about forty genera and more than five hundred species. They are found in both temperate and tropical regions of the World. Several species are grown as ornamentals, and a few of these have escaped from cultivation in Illinois.

KEY TO THE GENERA OF Celastraceae IN ILLINOIS

1. Leaves alternate; most of the flowers unisexual _____ 1. *Celastrus*
1. Leaves opposite; flowers perfect_____ 2. *Euonymus*

1. *Celastrus* L. –Bittersweet

Woody vines; leaves alternate, simple; flowers in terminal panicles or in axillary cymes, small, perfect or unisexual; calyx small, 5-parted; petals 5, free; stamens 5, attached to a disk, rudimentary in pistillate flowers; ovary superior, rudimentary in staminate flowers; stigma 3-lobed; fruit a 3-valved capsule; seeds covered by a red aril.

Celastrus is a genus of about thirty species, most of them native to eastern Asia. One native and one introduced species occur in Illinois.

KEY TO THE SPECIES OF Celastrus IN ILLINOIS

1. Flowers in terminal panicles or racemes; leaves ovate to ovate-oblong, finely serrate _____ 1. *C. scandens*

1. Flowers in axillary cymes; leaves suborbicular, crenate _____
_____ 2. *C. orbiculatus*

1. Celastrus scandens L. Sp. Pl. 196. 1753. *Fig. 90.*

Woody climber, sometimes growing to the tops of trees; leaves
ovate to ovate-oblong, acute to acuminate at the apex, cuneate to
the base, finely serrate, glabrous, to 10 cm long, to 4 cm broad;
flowers in terminal panicles or racemes; petals greenish, 2–3 mm
long; fruits globose, orange-yellow to orange, up to 1 cm in diameter; seeds covered by a bright red aril, subglobose to ellipsoid, up
to 8 mm in diameter.

COMMON NAME: Bittersweet.
HABITAT: Woods and thickets.
RANGE: Quebec to Manitoba, south to Wyoming,
Texas, Louisiana, and Georgia.
ILLINOIS DISTRIBUTION: Occasional throughout the
state.
Bittersweet is one of the most attractive and, as a re-
sult, one of the most sought after vines in the Illinois
flora.
Its brilliant red arillate seeds are exposed during Octo-
ber when the surrounding orange-yellow valves of the
capsule split open.

This species occurs in most kinds of woods, from low woods near
rivers and streams to dry upland woods.

The flowers bloom in May and June.

2. Celastrus orbiculatus Thunb. Fl. Jap. 42. 1784. *Fig. 91.*

Celastrus articulatus Thunb. Fl. Jap. 97. 1784.

Woody climber; leaves suborbicular to broadly obovate, acute to
acuminate at the apex, cuneate to rounded at the base, crenate,
glabrous, to 10 cm long, to 6 cm broad; flowers in axillary cymes;
petals greenish, 2–3 mm long; fruits globose, orange-yellow to
orange, up to 1 cm in diameter; seeds covered by a brilliant red
aril, usually subglobose, up to 8 mm in diameter.

90. *Celastrus scandens* (Bittersweet). *a.* Habit, in flower, ×½. *b.* Habit, in fruit, ×½. *c.* Staminate flower, ×5. *d.* Perfect flower, ×5. *e.* Seed, ×2½.

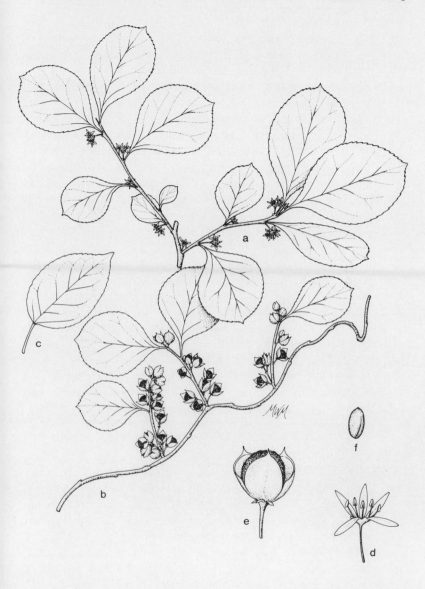

91. *Celastrus orbiculatus* (Round-leaved Bittersweet). *a.* Habit, with staminate flowers, ×½. *b.* Habit, with fruits, ×½. *c.* Leaf variation, ×½. *d.* Staminate flower, ×2½. *e.* Fruit, ×2½. *f.* Seed, ×2½.

COMMON NAME: Round-leaved Bittersweet.

HABITAT: Thickets.

RANGE: Native of Asia; occasionally escaped from cultivation in the eastern United States.

ILLINOIS DISTRIBUTION: Rarely escaped but scattered in Illinois.

This species is occasionally grown as an ornamental. The first escape from cultivation in Illinois was found at Mississippi Palisades State Park in Carroll County on May 17, 1941.

It is relatively common in Giant City State Park in Jackson and Union counties, where it is beginning to overrun some native habitats.

The flowers bloom in May and June.

2. *Euonymus* L. –Spindle Tree

Small trees, shrubs, or climbers, rooting at the nodes; leaves opposite, simple, sometimes evergreen; flower solitary or in small axillary clusters, perfect; sepals 4 or 5, green, united at the base; petals 4 or 5, free; stamens 5, attached to a disk, the disk covering the ovary; ovary superior; stigma 3- to 5-lobed; fruit a 3-valved capsule; seeds covered by a fleshy red aril.

Several of the one hundred twenty species of *Euonymus* are popular ornamentals, and some of these have been found as escapes in Illinois.

KEY TO THE SPECIES OF Euonymus IN ILLINOIS

1. Plants trailing or climbing, rooting at the nodes _____ 2
1. Plants erect or ascending, not rooting at the nodes_____ 4
    2. Leaves membranaceous, deciduous, obovate to oblong; sepals and petals each 5; capsule warty; plants trailing _____ 1. *E. obovata*
    2. Leaves coriaceous, sometimes evergreen, ovate or elliptic; sepals and petals each 4; capsule smooth; plants climbing _____ 3
3. Leaves ovate, to 6 cm long; flowers in short-peduncled cymes _____
    _____ 2. *E. kiautschovica*
3. Leaves elliptic, to 4 cm long; flowers in long-peduncled cymes _____
    _____ 3. *E. fortunei*
    4. Stems unwinged _____ 5
    4. Stems winged _____ 7. *E. alata*
5. Leaves petiolate, the petioles mostly more than 3 mm long; sepals and petals each 4; capsule smooth; shrubs to 7 m tall _____ 6

5. Leaves sessile, or on petioles up to 3 mm long; sepals and petals each 5; capsule warty; shrubs to 2.5 m tall _____ 6. *E. americana*
6. Leaves pubescent on the lower surface; flowers purple _____ _____ 4. *E. atropurpurea*
6. Leaves glabrous on the lower surface; flowers yellow-green_____ _____ 5. *E. europaea*

1. **Euonymus obovata** Nutt. Gen. 1:155. 1818. *Fig. 92.*
Euonymus americanus L. var. *obovatus* (Nutt.) Torr. & Gray, Fl. N. Am. 1:258. 1838.

Prostrate shrub, rooting at the nodes, the stems usually glabrous; leaves deciduous, obovate to oblong, acute to obtuse at the apex, cuneate to the base, serrulate, glabrous, to 6 cm long, to 3.5 cm broad, on glabrous petioles up to 5 mm long; flowers axillary, solitary or in few-flowered cymes, on pedicels to 8 mm long; sepals 5, green, very small; petals 5, greenish purple, 2–4 mm long, suborbicular, not clawed; disk broad, flat, 5-angled, concealing the ovary; stamens 5, borne at the edge of the disk; fruit usually 3-lobed, warty, up to 1.5 cm in diameter; seeds with a scarlet aril.

COMMON NAME: Running Strawberry Bush.
HABITAT: Rich woods.
RANGE: Ontario and New York to Michigan, south to Missouri, Tennessee, and West Virginia.
ILLINOIS DISTRIBUTION: Scattered but uncommon throughout the state.
This species is similar to *E. americana* and has been confused with it by some of the earlier Illinois collectors.
The most obvious difference between the two species is the growth form, with the stems of *E. obovata* prostrate at first, while those of *E. americana* are all erect.

The scarlet aril that surrounds each seed is exposed when the capsule dehisces.

Euonymus obovata flowers from late April to early June.

2. **Euonymus kiautschovica** Loes. Bot. Jahrb. 30:453. 1902. *Fig. 93.*

Climbing shrub with aerial roots; leaves evergreen, coriaceous, ovate, acute at the apex, rounded to cuneate at the base, serrulate, glabrous, to 6 cm long, to 4 cm broad; flowers axillary, borne in

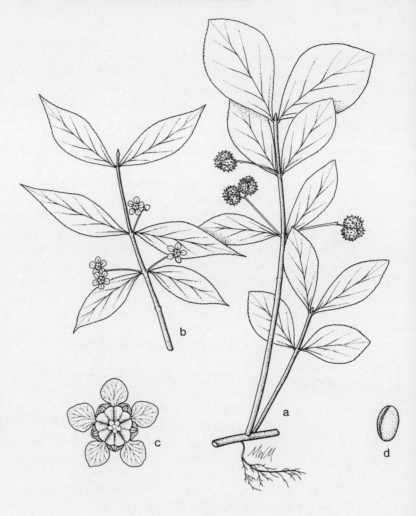

92. *Euonymus obovata* (Running Strawberry Bush). *a*. Habit, in fruit, × ½. *b*. Flowering branch, × ½. *c*. Flower, × 2½. *d*. Seed, × 2½.

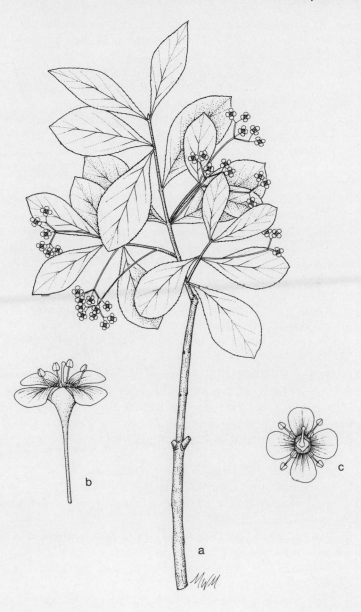

93. Euonymus kiautschovica (Climbing Euonymus). *a*. Habit, ×½. *b*. Flower, side-view, ×3. *c*. Flower, face-view, ×2½.

cymes on short peduncles; sepals 4, green, very small; petals 4, greenish white, 2–3 mm long, disk present; fruit globose, not warty; seeds arillate.

COMMON NAME: Climbing Euonymus.

HABITAT: Climbing on a tree (in Illinois).

RANGE: Native to Asia; rarely escaped from cultivation in the United States.

ILLINOIS DISTRIBUTION: Known only from Jackson County, where a specimen was found growing adventively on a tree in Murphysboro.

This species is uncommon as an ornamental in the Midwest, and it rarely escapes from cultivation.

Euonymus kiautschovica is very similar to the more commonly grown *E. fortunei*, but differs by its larger, ovate leaves and its short-pedunculate cymes.

The flowers bloom during July and August.

3. **Euonymus fortunei** (Turcz.) Hand.-Maz. Symb. Sin. 7:660.1913. *Fig. 94.*

Elaeodendron fortunei Turcz. Bull. Soc. Nat. Mosc. 36(1):603. 1866.

Climbing shrub with aerial roots; leaves evergreen, coriaceous, elliptic, acute at the apex, cuneate to the base, serrulate, glabrous, to 4 cm long, to 1.5 cm broad; flowers axillary, borne in cymes on long peduncles; sepals 4, green, very small; petals 4, greenish white, 2–3 mm long; disk present; fruit globose, not warty; seeds arillate.

COMMON NAME: Climbing Euonymus.

HABITAT: Disturbed areas.

RANGE: Native of Asia; occasionally adventive in the United States.

ILLINOIS DISTRIBUTION: Rare and scattered in the state.

This is a rather frequently planted ornamental in Illinois.

At Giant City State Park, *Euonymus fortunei* has become firmly established and is spreading rapidly in a low woods.

The flowers bloom in July and August.

94. Euonymus fortunei (Climbing Euonymus). *a.* Habit, × ½.

4. **Euonymus atropurpurea** Jacq. Hort. Vind. 2:5. 1772 *Fig. 95.*

Small tree or shrub to 7 meters; stems glabrous, not 4-angled, the young twigs usually greenish; leaves deciduous, elliptic to narrowly ovate, acute to acuminate at the apex, rounded to cuneate at the base, serrulate, glabrous on the upper surface, pubescent on the lower surface, to 12 cm long, to 6 cm broad, petiolate, the petioles at least 3 mm long; flowers in axillary cymes on peduncles longer than the petioles; sepals 4, green; petals 4, purple, 3–4 mm long; disk present; fruit irregularly 4-lobed, pink to rose, smooth, up to 1.5 cm in diameter, some of the lobes usually with aborted seeds; fertile seeds with a bright red aril.

COMMON NAME: Wahoo.

HABITAT: Woods.

RANGE: Ontario to Montana, south to Texas and Florida.

ILLINOIS DISTRIBUTION: Occasional throughout the state.

Wahoo is one of the more attractive small trees in Illinois. The leaves often turn red in the autumn, and the rose-colored fruits expose the bright red arils of the seeds on dehiscence.

The purple flowers open during June and July.

5. **Euonymus europaea** L. Sp. Pl. 197. 1753 *Fig. 96.*

Small tree to 6 meters; stems glabrous, not 4-angled; leaves deciduous, elliptic to broadly lanceolate, acuminate at the apex, cuneate to the base, serrulate, glabrous on both surfaces, to 8 cm long, to 3 cm broad, petiolate, the petioles at least 3 mm long; flowers in axillary cymes on peduncles longer than the petioles; sepals 4, green; petals 4, yellow-green, 3–5 mm long; disk present; fruit 4-lobed, pink, up to 1.5 cm in diameter; seeds with an orange aril.

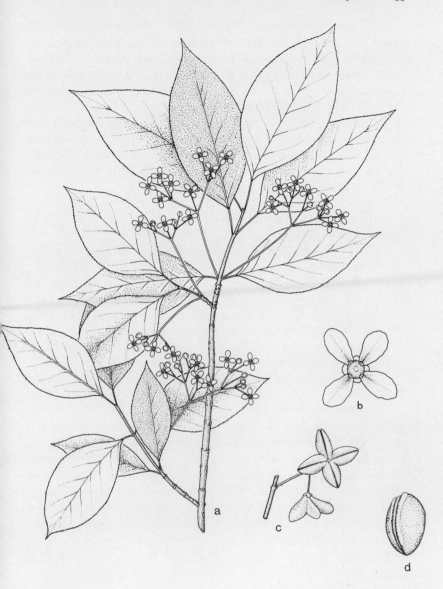

95. *Euonymus atropurpurea* (Wahoo). *a.* Habit, ×½. *b.* Flower, ×2½. *c.* Fruits, ×1. *d.* Seed, ×2½.

96. *Euonymus europaea* (European Spindle-tree). *a.* Habit, × ½. *b.* Flowering branch, × ½. *c.* Flower, × 3.

COMMON NAME: European Spindle-tree.

HABITAT: Floodplains (in Illinois).

RANGE: Native of Europe; occasionally escaped from cultivation in the United States.

ILLINOIS DISTRIBUTION: Known from stations in Du Page and Kane counties.

The European spindle-tree resembles the wahoo in general appearance, but differs by its narrower leaves, fewer-flowered cymes, yellow-green flowers, and orange arils. It is more commonly seen as an ornamental.

The flowers bloom during May and June.

6. Euonymus americana L. Sp. Pl. 197. 1753. *Fig. 97.*

Small erect shrub to 2.5 meters; stems 4-angled, glabrous; leaves deciduous, lanceolate to ovate, acute to acuminate at the apex, cuneate to the base, serrulate, glabrous on both surfaces, to 6 cm long, to 4 cm broad, sessile or on petioles up to 3 mm long; flowers axillary, solitary or 2–3 in a cyme, on long peduncles; sepals 5, green; petals 5, greenish purple, nearly orbicular except for the claw, 5–6 mm long; disk present; fruit 3- to 5-lobed, red, warty, up to 1.5 cm in diameter; arils scarlet.

COMMON NAME: Strawberry-bush.

HABITAT: Rich woods.

RANGE: New York to Missouri, south to Texas and Florida.

ILLINOIS DISTRIBUTION: Known from Vermilion and Wabash counties and five counties in the extreme southern end of the state.

This small, erect shrub is sometimes confused with the prostrate *Euonymus obovata*. *Euonymus americana* is further distinguished by its clawed petals.

The first Illinois collection of this plant was made by Jacob Schneck in Wabash County during the latter part of the nineteenth century.

The time of flowering for this species is May and June.

7. Euonymus alata (Thunb.) Sieb. Verh. Batav. Genoot. Kurst. Wetensch. 12:49. 1830. *Fig. 98.*
Celastrus alatus Thunb. Fl. Jap. 98. 1784.

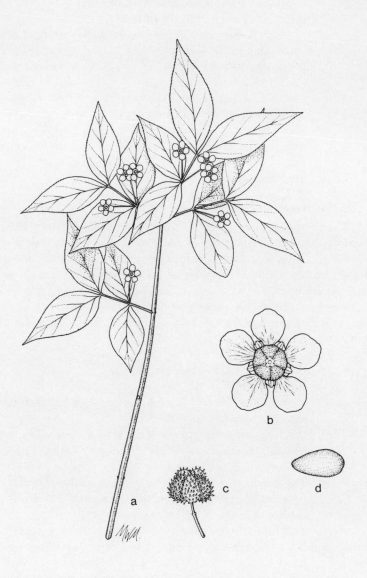

97. *Euonymus americana* (Strawberry-bush). *a.* Habit, ×½. *b.* Flower, ×2½. *c.* Fruit, ×1. *d.* Seed (without aril), ×3.

98. *Euonymus alata* (Winged Euonymus; Burning Bush). *a.* Habit, ×½. *b.* Flower, ×4. *c.* Paired fruits, ×2½. *d.* Seed (without aril), ×3.

Usually bushy-branched shrub to 3 meters tall; stems with flattened corky wings on either side; leaves deciduous, broadly elliptic, acute at the apex, cuneate to the base, serrulate, glabrous on both surfaces, to 4 cm long, to 2 cm broad; flowers axillary in cymes; sepals 4, green; petals 4, greenish yellow, 2–3 mm long; disk present; fruit purple, smooth; aril orange.

COMMON NAMES: Winged Euonymus; Burning Bush.
HABITAT: Disturbed areas.
RANGE: Native of eastern Asia; occasionally adventive in the United States.
ILLINOIS DISTRIBUTION: Scattered in Illinois.
Euonymus alata is a highly prized ornamental in Illinois because of its globular growth form and the brilliant autumnal coloration. These characters provide the basis for the common name of burning bush. The corky winged stems are another distinctive feature of this species.
The flowers bloom during June and July.

SANTALACEAE–SANDALWOOD FAMILY

Herbs (in Illinois), shrubs, or trees, often root parasites; leaves simple, alternate or opposite; flowers actinomorphic, perfect or unisexual, borne in axillary or terminal inflorescences or solitary; calyx 4- or 5-parted, united; petals absent; disk present; stamens 4 or 5, usually attached to the base of the calyx lobes; ovary inferior, 1-locular; fruit a drupe or nut, 1-seeded.

The sandalwood family is composed of about thirty genera and more than four hundred species, most of them in the tropics of the southern hemisphere. Several of these tropical species have considerable economic importance.

In Illinois, only the following genus represents this family.

1. *Comandra* Nutt.–Bastard Toadflax

Rhizomatous herbs, usually parasitic on the roots of other plants; leaves alternate, entire; flowers perfect, borne in cymes or panicles of small umbels; calyx 5-parted, united below; petals absent; disk 5-lobed; stamens 5, connected to the base of the calyx lobes by a tuft of hairs; ovary inferior, 1-locular; fruit a hard drupe (sometimes considered a nut) enclosed at the base by the persistent calyx, 1-seeded.

Of the six species that are in this genus, one is European, two are in the western United States, one is in the plains of the United States, and two are in the eastern United States.

Only the following species occurs in Illinois.

1. **Comandra umbellata** (L.) Nutt. Gen. N. Am. Pl. 1:157. 1818. *Fig.* 99.

Thesium umbellatum L. Sp. Pl. 208. 1753.

Comandra richardsiana Fern. Rhodora 7:48. 1905.

Herb with slender rhizomes at or just below the surface of the soil; aerial stems erect, slender, to 30 cm tall, glabrous; leaves oblong to lance-ovate, obtuse to acute at the apex, cuneate to the base, entire, glabrous, to 4 cm long, to 2.5 cm broad, sessile or nearly so; flowers in small, usually terminal, flat-topped clusters; calyx 5-parted, the lobes 2–3 mm long, whitish; petals absent; disk shallowly lobed; stamens 5, opposite the lobes of the calyx; drupe dry, subglobose, to 6 mm in diameter.

COMMON NAME: Bastard Toadflax.

HABITAT: Woods, prairies.

RANGE: Newfoundland to Quebec, south to Kansas, Arkansas, and New York.

ILLINOIS DISTRIBUTION: Occasional throughout the state.

An interesting feature about this species and most other species in the family is the association with other plants as root parasites. *Comandra richardsiana* parasitizes the roots of several different kinds of trees.

Although some botanists consider some or all of our specimens to be *C. richardsiana,* I am following Piehl (1965) who is unable to distinguish *C. umbellata* and *C. richardsiana.*

The flowers bloom from May to August.

VISCACEAE–MISTLETOE FAMILY

Green parasitic herbs or shrubs, parasitizing the aerial portions of a wide variety of woody hosts; leaves simple, opposite; flowers actinomorphic, perfect or unisexual, borne in various kinds of inflorescences; calyx 2- to 5-parted, usually united below; petals absent,

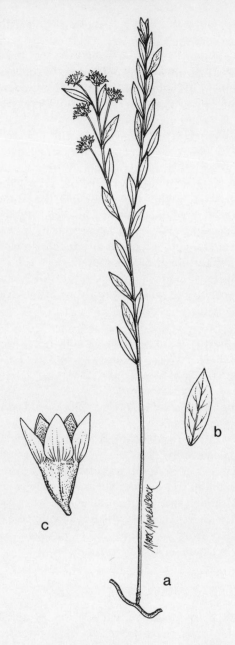

99. *Comandra umbellata* (Bastard Toadflax). *a.* Habit, × ½. *b.* Leaf, × 1. *c.* Flower, × 5.

except in some tropical species; stamens 2–5, usually attached to the calyx; ovary inferior, 1-locular; ovules mostly undifferentiated; fruit a berry or drupe.

Although *Phoradendron*, the only genus to occur in Illinois, traditionally has been placed in the Loranthaceae, several recent botanists have recognized Viscaceae as a family segregate of Loranthaceae and placed *Phorandendron* there.

This is mostly a tropical family occurring in both the New and Old Worlds. All the species are parasitic, attaching themselves to their host by means of haustoria. Some of the tropical species have showy flowers with well developed petals. The lone species that grows in Illinois lacks showy flowers and does not have petals. There are more than thirteen hundred species in this family.

Only the following genus occurs in Illinois.

1. *Phoradendron* Nutt.–Mistletoe

Parasitic shrub; leaves simple, opposite, entire; flowers unisexual, the sexes borne on separate plants, crowded in short, axillary spikes; calyx 3-parted, united below; petals absent; stamens 3, without filaments; ovary inferior, 1-locular; fruit a berry.

Phoradendron is a genus of more than three hundred species found mostly in the American tropics and subtropics.

Only the following species occurs in Illinois.

1. **Phoradendron serotinum** (Raf.) M. C. Johnst. Southw. Nat. 2:45. 1957. *Fig. 100.*
Viscum flavescens Pursh, Fl. Am. Sept. 114. 1814.
Viscum serotinum Raf. Ann. Gen. Sci. Phys. 5:348. 1820.
Phoradendron flavescens (Pursh) Nutt. ex Gray, Man. Bot., ed. 2, 383. 1856.

Parasitic shrubs, usually growing on trees or shrubs; stems glabrous, usually brittle, branched; leaves oblong to oblanceolate, broadly rounded at the apex, cuneate to the base, entire, glabrous, to 5 cm long, to 2.5 cm broad; spikes 1 (–2) in the axils of the leaves, to 5 cm long; flowers inconspicuous, greenish white, 2–3 mm long; berry subglobose, white, sticky, to 5 mm in diameter.

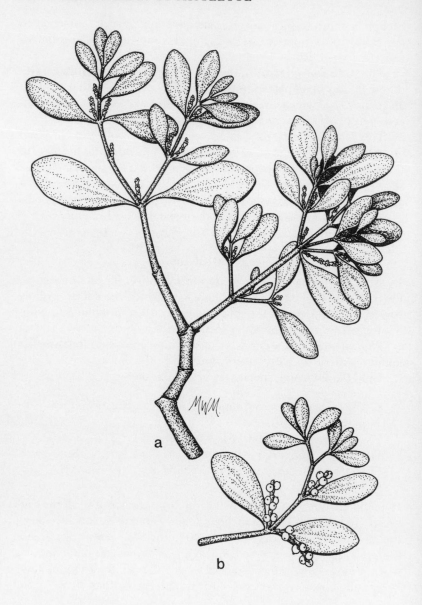

100. Phoradendron serotinum (Mistletoe). *a.* Branchlets, with flowers, × ½. *b.* Branchlet, with fruits, × ½.

COMMON NAME: Mistletoe.

HABITAT: Parasitic on trees or shrubs, mostly in low woods.

RANGE: New Jersey to Missouri, south to Texas and Florida; New Mexico.

ILLINOIS DISTRIBUTION: Occasionally in the southern one-sixth of the state, but extending along the Wabash River to Clark County.

Mistletoe is a remarkable plant in the Illinois flora. Although possessing chlorophyll in its leaves that enables it to manufacture food, it also obtains nutrients from the host species it parasitizes. The hosts in southern Illinois for the mistletoe include American elm and sweet gum. Excessive numbers of mistletoe plants on a single host may ultimately result in the death of the host.

Mistletoe seeds are dissiminated by birds that attempt to eat the berries. The sticky berries adhere to the bird's beak. In order to rid itself of the berry, the bird will rub its beak back and forth on the branch of a tree, finally transferring the berry from beak to branch.

Because of its popularity at Christmastime as a traditional decoration, mistletoe is sought after and is becoming less common in Illinois.

The flowers bloom during September and October.

Species Excluded

Convolvulus macrorhizos L. Mead (1846) used this binomial erroneously for *C. silvatica* var. *fraterniflorus*.

Cuscuta umbrosa Hook. Lapham (1857) used this binomial for *C. gronovii*. *Cuscuta umbrosa* does not occur in Illinois.

Datura metel L. This species does not occur in Illinois, although the binomial was used by a few earlier Illinois botanists for *D. innoxia*.

Ipomoea nil (L.) Roth, attributed to Illinois by several early botanists, is a species that lives south of Illinois. These early workers were actually referring to *I. hederacea*.

Physalis lanceifolia Nees. Rydberg (1896) used this binomial for *P. pendula*. *Physalis lanceifolia* does not occur in Illinois.

Physalis pennsylvanica L. Although Engelmann (1843), Mead (1846), Babcock (1872), Patterson (1874; 1876), and Seymour (1883) all listed this binomial for Illinois, their references were actually for *P. virginiana*.

Physalis philadelphica Lam. Several early Illinois botanists used this binomial erroneously for *P. subglabrata*.

Physalis viscosa L. This Linnaean binomial was attributed by several early Illinois botanists to either *P. pruinosa* or *P. heterophylla*. *Physalis viscosa* does not occur in Illinois.

Phlox amoena Sims. This species, from the eastern and southern United States, was erroneously used for *P. divaricata* by Huett (1897).

Phlox procumbens Lehm. Schneck (1876) erroneously used this binomial for *P. divaricata*.

Phlox pyramidalis Sm. This binomial, used by Mead (1846) for *P. divaricata*, is a species unknown from Illinois.

Phlox reptans Michx. The use of *P. reptans* for Illinois plants by several botanists during the nineteenth century was in error for *P. divaricata*.

Solanum atropurpureum Schrank. This binomial was used erroneously by Kibbe (1952) for *S. rostratum*.

Solanum nigrum L. Although this binomial has been in use in Illinois for the black nightshade, it actually refers to a species not known from Illinois. Our black nightshade is *S. ptycanthum*.

Summary of the Taxa Treated in This Volume

Families	Genera	Species	Lesser Taxa	Hybrids
Solanaceae	9	36	3	1
Convolvulaceae	6	14	2	
Cuscutaceae	1	10	2	
Polemoniaceae	6	13	4	
Campanulaceae	3	14	3	1
Celastraceae	2	9		
Santalaceae	1	1		
Viscaceae	1	1	—	—
Totals	29	98	14	2

GLOSSARY
LITERATURE CITED
INDEX OF PLANT NAMES

GLOSSARY

Actinomorphic. Having radial symmetry; regular, in reference to a flower.

Acuminate. Gradually tapering to a point.

Acute. Sharply tapering to a point.

Annual. Living for only one year.

Anther. The terminal part of a stamen that bears pollen.

Anthesis. Flowering time.

Apiculate. Abruptly short-pointed at the tip.

Appendage. An accessory part attached to a structure.

Appressed. Lying flat against the surface.

Aril. An appendage of the seed, usually enclosing the seed.

Arillate. Bearing an aril.

Articulated. Jointed.

Attenuate. Gradually becoming narrowed.

Auricle. An earlike lobe.

Auriculate. Bearing an earlike process.

Axil. The angle where two structures join.

Axile. Said of ovules that are borne at or near the center of a compound ovary.

Beak. A slender terminal projection.

Bearded. Bearing a tuft of stiff hairs.

Berry. A type of fruit where the seeds are surrounded only by fleshy material.

Biennial. Completing a life cycle in two years.

Bifid. Two-cleft.

Bisexual. Referring to a flower that contains both stamens and pistils.

Bract. An accessory structure at the base of a flower, usually appearing leaflike.

Bracteate. Bearing one or more bracts.

Bracteole. A secondary bract.

Calyx. The outermost segments of the perianth of a flower, composed of sepals.

Campanulate. Bell-shaped.

Canescent. Gray-hairy.

Capitate. Forming a head.

Capsule. A dry, dehiscent fruit composed of more than one carpel.

Carinate. Having a ridge or keel.

Carpel. A simple pistil, or one member of a compound pistil.

Cauline. Belonging to a stem.

Chlorophyll. The green pigment in a plant.

Cilia. Marginal hairs.

Ciliate. Bearing cilia.

Cinereous. Ashy gray.

Circumscissile. Usually referring to a fruit that dehisces by a horizontal, circular line.

Clasping. Projecting around a structure, such as the base of a leaf around a stem.

Cleistogamous. Hidden.

Compound. Divided into more than one segment or part.

Conical. Cone-shaped.

Connivent. Coming together.

Cordate. Heart-shaped.

Coriaceous. Leathery.

Corolla. The segments of a flower just within the calyx, composed of petals.

Corymb. A type of inflorescence where the pedicellate flowers are arranged along an elongated axis, but with the flowers all attaining about the same height.

Crenate. With round teeth.

Crenulate. With small, round teeth.

Cuneate. Wedge-shaped.

Cuspidate. Terminating in a very short point.

Cylindrical. Shaped like a cylinder.

Cyme. A type of broad and flattened inflorescence in which the central flowers bloom first.

Cymose. Bearing a cyme.

Deciduous. Falling away.

Decumbent. Lying flat, but with the tip ascending.

Decurrent. Attached to the petiole or stem and then extending beyond the point of attachment.

Dehiscent. Splitting at maturity.

Deltoid. Triangular.

Dentate. With sharp teeth, the tips of which project outward.

Denticulate. With small, sharp teeth, the tips of which project outward.

Depressed. Pushed down flat along the summit.

Diffuse. Loosely spreading.

Disk. An enlarged outgrowth of the receptacle.

Divergent. Spreading apart.

Drupe. A type of fruit in which the seed is surrounded by a hard, dry covering which, in turn, is surrounded by fleshy material.

Ellipsoid. Referring to a solid object that is broadest at the middle, gradually tapering to both ends.

Elliptic. Broadest at middle, gradually tapering to both ends.

Emarginate. Having a shallow notch at the extremity.

Entire. Without teeth.

Exserted. Protruded beyond.

Fascicle. Cluster.

Fibrous. Referring to roots borne in tufts.

Filament. That part of the stamen supporting the anther.

Filiform. Threadlike.

Fimbriate. Fringed.

Flexuous. Zigzag.

Foliaceous. Leaflike.

Funnelform. Shaped like a funnel.

Glabrous. Without pubescence or hairs.

Gland. An enlarged, usually spherical body functioning as a secretory organ.

Glandular. Bearing glands.

Globose. Round.

Glomerule. A small, compact cluster.

Hastate. Spear-shaped; said of a leaf that is triangular, with spreading basal lobes.

Haustoria. Adhesivelike disks

that attach a parasitic plant to its host.

Head. A usually spherical, compact cluster.

Herbaceous. Not woody.

Hirsute. With stiff hairs.

Hirsutulous. With minute stiff hairs.

Hirtellous. Finely hirsute.

Hispid. With rigid hairs.

Hispidulous. With minute rigid hairs.

Hoary. Grayish white.

Included. Nor protruding.

Inferior. Referring to the position of the ovary when it is surrounded by the attached portion of the floral tube or is embedded in the receptacle.

Inflexed. Turned inward.

Inflorescence. A cluster of flowers.

Internode. The area between two adjacent nodes.

Keel. A ridge.

Lacerated. Divided into shreds.

Lanceolate. Lance-shaped; broadest near base, gradually tapering to the narrower apex.

Latex. Milky juice.

Leaflet. An individual unit of a compound leaf.

Lenticular. Lens-shaped.

Linear. Elongated and uniform in width throughout.

Locular. Referring to the locule, or cavity of the ovary.

Locule. The cavity of an ovary.

Lustrous. Shiny.

Membranaceous. Thin and transparent.

Mucronate. Possessing a short, abrupt tip.

Node. That place on the stem from which leaves and branchlets arise.

Oblanceolate. Reverse lance-shaped; broadest at apex, gradually tapering to narrow base.

Oblong. Broadest at the middle, and tapering to both ends, but broader than elliptic.

Oblongoid. Referring to a solid object that, in side view, is nearly the same width throughout.

Obovate. Reverse egg-shaped; broadly rounded at apex, becoming narrowed below.

Obsolete. Almost absent.

Obtuse. Rounded at the apex.

Occluded. Closed or nearly so.

Opaque. Incapable of being seen through.

Orbicular. Round.

Oval. Broadly elliptic.

Ovary. The lower swollen part of the pistil that produces the ovules.

Ovate. Broadly rounded at base, becoming narrowed above.

Ovule. An immature seed.

Panicle. A type of inflorescence composed of several racemes.

Paniculate. Bearing panicles.

Pedicel. The stalk of a flower in an inflorescence.

Pedicellate. Bearing a pedicel.

Peduncle. The stalk of an inflorescence.

Pendulous. Hanging.

Perennial. Living more than two years.

Perfect. Bearing both stamens and pistils in the same flower.

Perianth. Those parts of a flower including both the calyx and corolla.

Petiolate. Bearing a petiole.

Petiole. The stalk of a leaf.

Pilose. Bearing soft hairs.

Pinnate. Divided once into distinct segments.

Pinnatifid. Said of a simple leaf or leaf-part that is cleft or lobed only partway to its axis.

Pinnatisect. Very deeply pinnately divided.

Pistillate. Bearing pistils but not stamens.

Placenta. That place in an ovary to which the ovules are attached.

Pore. A small, circular opening.

Prostrate. Lying flat.

Puberulent. With minute hairs.

Pubescent. Bearing some kind of hairs.

Pyramidal. Shaped like a pyramid.

Raceme. A type of inflorescence where pedicellate flowers are arranged along an elongated axis.

Racemose. Bearing racemes.

Receptacle. That part of the flower to which the perianth, stamens, and pistils are usually attached.

Reclining. Leaning on an object.

Reflexed. Turned downward.

Repand. Wavy along the margins.

Reticulate. Resembling a network.

Retrorse. Pointing downward.

Rhizomatous. Bearing rhizomes.

Rhizome. An underground, horizontal stem bearing nodes, buds, and roots.

Rosette. A cluster of leaves in a circular arrangement at the base of a plant.

Rotate. Flat and circular.

Rugose. Wrinkled.

Saccate. Sac-shaped, or pouchlike.

Sagittate. Shaped like an arrowhead.

Salverform. Referring to a tubular corolla that abruptly expands into a flat limb.

Scabrous. Rough to the touch.

Secund. Borne on one side.

Serrate. With teeth that project forward.

Serrulate. With very small teeth that project forward.

Sessile. Without a stalk.

Simple. Said of a leaf that is not divided into leaflets.

Sinuate. Wavy along the margins.

Sinus. The cleft between two lobes or teeth.

Spatulate. Oblong, but with the basal end elongated.

Spherical. Circular.

Spike. A type of inflorescence where sessile flowers are arranged along an elongated axis.

Stamen. The pollen-producing

organ of a flower composed of a filament and an anther.

Staminate. Bearing stamens but not pistils.

Stellate. Star-shaped.

Stigma. That part of the pistil that receives pollen.

Stipitate. Possessing a small stalk.

Stipule. A leaflike or scaly structure found at the point of attachment of a leaf to the stem.

Striate. Marked with grooves.

Strigose. With appressed, straight hairs.

Style. The elongated part of the pistil between the stigma and the ovary.

Stylopodium. A small base to a structure.

Subacute. Nearly pointed at the tip.

Subapical. Almost near the tip.

Subcordate. Nearly heart-shaped.

Subcuneate. Nearly wedge-shaped.

Suborbicular. Nearly spherical.

Subrotate. Nearly flat and spreading.

Subulate. With a very short, narrow point.

Sucker. A structure that enables a parasitic plant to adhere to its host.

Superior. Referring to the position of the ovary when the free floral parts arise below the ovary.

Tawny. Light brown.

Thyrse. A type of inflorescence in which the main axis is racemose and the lateral axes are cymose.

Tomentose. Pubescent with matted wool.

Tomentum. Woolly hair.

Translucent. Partly transparent.

Truncate. Abruptly cut across.

Tuber. An underground, fleshy stem found as a storage organ at the end of a rhizome.

Tubular. Elongated; shaped like a tube.

Turbinate. Top-shaped, like an inverted cone.

Umbel. A type of inflorescence in which the flower stalks arise from the same level.

Umbellate. Bearing umbels.

Undulate. Wavy.

Unisexual. Having either stamens or pistils in a flower, but not both.

Urceolate. Urn-shaped.

Valve. A segment of a capsule.

Villi. Long soft hairs.

Villous. With long, soft, slender, unmatted hairs.

Zygomorphic. Bilaterally symmetrical.

LITERATURE CITED

Babcock, H. H. 1872. The flora of Chicago and vicinity. Lens 1:20–26, 65–71, 144–50, 169, 218–22.

Brummitt, R. K. 1965. New combinations in North American *Calystegia*. Annals of the Missouri Botanical Garden 52:214–16.

———. 1980. Further new names in the genus *Calystegia*. Kew Bulletin 35:327–31.

Cronquist, A. 1959. *Campanula*. In C. L. Hitchcock, Vascular plants of the Pacific northwest. Seattle: Univ. of Washington Press. Pp. 458–59.

———. 1981. An integrated system of classification of flowering plants. New York: Columbia Univ. Press. 1,262 pp.

Engelmann, G. 1842. A monograph of North American Cuscutineae. American Journal of Science 43:333–45.

———. 1843. Catalogue of collection of plants made in Illinois and Missouri by Charles A. Geyer. American Journal of Science 46:94–104.

———. 1845. Additions to *Cuscuta*. Boston Journal of Natural History 5:224.

Fernald, M. L. 1950. Gray's manual of botany. 8th ed. New York: American Book Company. 1,632 pp.

Gleason, H. A. 1952. Change of name for certain plants of the "Manual Range." Phytologia 4:20–25.

Grant, V. 1956. A synopsis of *Ipomopsis*. Aliso 3:351–62.

Huett, J. W. 1897. Essay toward a natural history of La Salle County, Illinois. Flora La Sallensis. Pt. I. Peru, Ill.: Privately published.

Kartesz, J. T., & R. Kartesz. 1980. A synonymized checklist of the vascular flora of the United States, Canada, and Greenland. Chapel Hill: Univ. of North Carolina Press. 500 pp.

Kibbe, A. 1952. A botanical study and survey of a typical midwestern county. Carthage, Ill.: Privately published. 425 pp.

Lapham, I. A. 1857. Catalogue of the plants of the state of Illinois. Transactions of the Illinois State Agriculture Society 2:492–550.

Levin, D. A. 1963. Natural hybridization between *Phlox maculata* and *Phlox glaberrima* and its evolutionary significance. American Journal of Botany 50:714–20.

———. 1966. The *Phlox pilosa* complex: crossing and chromosome relationships. Brittonia 18:142–62.

———, & D. M. Smith. 1965. An enigmatic *Phlox* from Illinois. Brittonia 17:254–66.

Lewis, W. H., & R. L. Oliver. 1965. Realignment of *Calystegia* and *Convolvulus* (Convolvulaceae). Annals of the Missouri Botanical Garden 52:217–22.

McVaugh, R. 1936. Studies in the taxonomy and distribution of eastern North American species of *Lobelia*. Rhodora 38:243–63, 276–98, 305–29, 346–62.

———. 1945. The genus *Triodanis* Rafinesque, and its relationships to *Specularia* and *Campanula*. Wrightia 1:13–52.

Mead, S. B. 1846. Catalogue of plants growing spontaneously in the state of Illinois, the principal part near Augusta, Hancock County. Prairie Farmer 6:35–36, 60, 93, 119–22.

Mohlenbrock, R. H. 1986. Guide to the vascular flora of Illinois. Revised and enlarged edition. Carbondale: Southern Illinois Univ. Press. 508 pp.

Myint, T. 1966. Revision of the genus *Stylisma* (Convolvulaceae). Brittonia 18:99–117.

Patterson, H. N. 1874. A list of plants collected in the vicinity of Oquawka, Henderson County. Oquawka, Ill.: Privately published. 18 pp.

———. 1876. Catalogue of the phaenogamous and vascular cryptogamous plants of Illinois. Oquawka, Ill.: Privately published. 54 pp.

Pepoon, H. S. 1927. An annotated flora of the Chicago area. Bulletin of the Chicago Academy of Science 8:1–554.

Piehl, M. A. 1965. The natural history and taxonomy of *Comandra* (Santalaceae). Memoirs of the Torrey Botanical Club 22(1):1–97.

Rydberg. P. A. 1896. The North American species of *Physalis* and related genera. Memoirs of the Torrey Botanical Club 4:297–374.

Schilling, E. E. 1981. Systematics of *Solanum* Sect. Solanum (Solanaceae) in North America. Systematic Botany 6:172–85.

Schneck, J. 1876. Catalogue of the flora of the Wabash Valley. Annual Report of the Geological Survey of Indiana 7:504–79.

Seymour, A. B. 1883. Flora of Piatt County. In Emma C. Piatt, History of Piatt County. Chicago: Privately published. Pp. 100–117.

Shetler, S. G. 1963. A checklist and key to the species of *Campanula* native or commonly naturalized in North America. Rhodora 65:319–37.

Steyermark, J. A. 1960. Flora of Missouri. Ames: Iowa State Univ. Press. 1,728 pp.

Swink, F., & G. S. Wilhelm. 1979. Plants of the Chicago region. Lisle, Ill.: Morton Arboretum. 922 pp.

Thorne, R. F. 1968. Synopsis of a putatively phylogenetic classification of the flowering plants. Aliso 6:57–66.

Waterfall, U. T. 1958. A taxonomic study of the genus *Physalis* in North America north of Mexico. Rhodora 60:107–14, 128–42, 152–73.

Wherry, E. 1936. Miscellaneous eastern Polemoniaceae. Bartonia 18:52–59.

————. 1955. The genus *Phlox*. Morris Arboretum Monographs. Morris Arboretum of the University of Pennsylvania. 174 pp.

Yuncker, T. G. 1921. Revision of the North American and West Indian species of *Cuscuta*. Univ. of Illinois Biological Monographs 6:1–141.

————. 1932. The genus *Cuscuta*. Memoirs of the Torrey Botanical Club 18:113–331.

INDEX OF PLANT NAMES

Robert H. Mohlenbrock has been studying the Illinois flora since his junior year in high school in Murphysboro, a period spanning more than forty years. During that time he has published scores of scientific and popular articles about the plants of the state. In addition to the monumental and highly acclaimed Illustrated Flora of Illinois series, of which *Nightshades to Mistletoe* is the thirteenth volume, he has published *Guide to the Vascular Flora of Illinois: Revised and Enlarged Edition.* He has also published on United States national forests and endangered plants of North America, and is a monthly columnist for *Natural History* magazine. Since 1957, Dr. Mohlenbrock has been on the faculty of Southern Illinois University at Carbondale, where he has directed eighty-six students to the successful completion of their graduate degrees. In 1985, he received the prestigious rank of Distinguished Professor, and, in 1988, was named Outstanding Scholar of the University.